Standing Operating Procedures for Developing Acute Exposure Guideline Levels for Hazardous Chemicals

Subcommittee on Acute Exposure Guideline Levels

Committee on Toxicology

Board on Environmental Studies and Toxicology

Commission on Life Sciences

National Research Council

NATIONAL ACADEMY PRESS
Washington, D.C.

NATIONAL ACADEMY PRESS 2101 Constitution Ave., N.W. Washington, D.C. 20418

NOTICE: The project that is the subject of this report was approved by the Governing Board of the National Research Council, whose members are drawn from the councils of the National Academy of Sciences, the National Academy of Engineering, and the Institute of Medicine. The members of the committee responsible for the report were chosen for their special competences and with regard for appropriate balance.

This project was supported by Contract Nos. DAMD17-89-C-9086 and DAMD17-99-C-9049 between the National Academy of Sciences and the U.S. Army. Any opinions, findings, conclusions, or recommendations expressed in this publication are those of the author(s) and do not necessarily reflect the view of the organizations or agencies that provided support for this project.

International Standard Book Number 0-309-07553-X

Library of Congress Control Number: 2001090090

Additional copies of this report are available from:

National Academy Press
2101 Constitution Ave., NW
Box 285
Washington, DC 20055

800-624-6242
202-334-3313 (in the Washington metropolitan area)
http://www.nap.edu

Copyright 2001 by the National Academy of Sciences. All rights reserved.

Printed in the United States of America

THE NATIONAL ACADEMIES

National Academy of Sciences
National Academy of Engineering
Institute of Medicine
National Research Council

The **National Academy of Sciences** is a private, nonprofit, self-perpetuating society of distinguished scholars engaged in scientific and engineering research, dedicated to the furtherance of science and technology and to their use for the general welfare. Upon the authority of the charter granted to it by the Congress in 1863, the Academy has a mandate that requires it to advise the federal government on scientific and technical matters. Dr. Bruce M. Alberts is president of the National Academy of Sciences.

The **National Academy of Engineering** was established in 1964, under the charter of the National Academy of Sciences, as a parallel organization of outstanding engineers. It is autonomous in its administration and in the selection of its members, sharing with the National Academy of Sciences the responsibility for advising the federal government. The National Academy of Engineering also sponsors engineering programs aimed at meeting national needs, encourages education and research, and recognizes the superior achievements of engineers. Dr. William A. Wulf is president of the National Academy of Engineering.

The **Institute of Medicine** was established in 1970 by the National Academy of Sciences to secure the services of eminent members of appropriate professions in the examination of policy matters pertaining to the health of the public. The Institute acts under the responsibility given to the National Academy of Sciences by its congressional charter to be an adviser to the federal government and, upon its own initiative, to identify issues of medical care, research, and education. Dr. Kenneth I. Shine is president of the Institute of Medicine.

The **National Research Council** was organized by the National Academy of Sciences in 1916 to associate the broad community of science and technology with the Academy's purposes of furthering knowledge and advising the federal government. Functioning in accordance with general policies determined by the Academy, the Council has become the principal operating agency of both the National Academy of Sciences and the National Academy of Engineering in providing services to the government, the public, and the scientific and engineering communities. The Council is administered jointly by both Academies and the Institute of Medicine. Dr. Bruce M. Alberts and Dr. William A. Wulf are chairman and vice chairman, respectively, of the National Research Council.

SUBCOMMITTEE ON ACUTE EXPOSURE GUIDELINE LEVELS

DANIEL KREWSKI, *(Chair)*, University of Ottawa, Ottawa, Ontario
EDWARD C. BISHOP, Parsons Engineering Science, Inc., Fairfax, Virginia
JAMES V. BRUCKNER, University of Georgia, Athens
JOHN DOULL, University of Kansas Medical Center, Kansas City
DONALD E. GARDNER, Inhalation Toxicology Associates, Inc., Raleigh, North Carolina
DAVID W. GAYLOR, U.S. Food and Drug Administration, Jefferson, Arkansas
FLORENCE K. KINOSHITA, Hercules Incorporated, Wilmington, Delaware
STEPHEN U. LESTER, Center for Health, Environment and Justice, Falls Church, Virginia
HARIHARA MEHENDALE, University of Louisiana, Monroe
RICHARD B. SCHLESINGER, New York University School of Medicine, Tuxedo
CALVIN C. WILLHITE, State of California, Berkeley

Staff

KULBIR S. BAKSHI, Project Director
RUTH E. CROSSGROVE, Editor
AIDA NEEL, Administrative Assistant

Sponsor: **U.S. Department of Defense**

COMMITTEE ON TOXICOLOGY

BAILUS WALKER, JR. *(Chair)*, Howard University Medical Center and American Public Health Association, Washington, D.C.
MELVIN E. ANDERSEN, Colorado State University, Denver
GERMAINE M. BUCK, National Institute of Health, Washington, D.C.
ROBERT E. FORSTER II, University of Pennsylvania, Philadelphia
WILLIAM E. HALPERIN, National Institute for Occupational Safety and Health, Cincinnati, Ohio
CHARLES H. HOBBS, Lovelace Respiratory Research Institute and Lovelace Biomedical and Environmental Research Institute, Albuquerque, New Mexico
SAM KACEW, University of Ottawa, Ottawa, Ontario
NANCY KERKVLIET, Oregon State University, Corvallis
MICHAEL J. KOSNETT, University of Colorado Health Sciences Center, Denver
MORTON LIPPMANN, New York University School of Medicine, Tuxedo
ERNEST E. MCCONNELL, ToxPath, Inc., Raleigh, North Carolina
THOMAS E. MCKONE, Lawrence Berkeley National Laboratory and University of California, Berkeley
HARIHARA MEHENDALE, University of Louisiana, Monroe
DAVID H. MOORE, Battelle Memorial Institute, Bel Air, Maryland
LAUREN ZEISE, California Environmental Protection Agency, Oakland

Staff

KULBIR S. BAKSHI, Program Director
SUSAN N.J. MARTEL, Program Officer
ABIGAIL E. STACK, Program Officer
RUTH E. CROSSGROVE, Publications Manager
AIDA NEEL, Administrative Assistant
JESSICA BROCK, Project Assistant

BOARD ON ENVIRONMENTAL STUDIES AND TOXICOLOGY

GORDON ORIANS (*Chair*), University of Washington, Seattle
JOHN DOULL, University of Kansas Medical Center, Kansas City
DAVID ALLEN, University of Texas, Austin
INGRID C. BURKE, Colorado State University, Fort Collins
THOMAS BURKE, Johns Hopkins University, Baltimore, Maryland
GLEN R. CASS, Georgia Institute of Technology, Atlanta
WILLIAM L. CHAMEIDES, Georgia Institute of Technology, Atlanta
CHRISTOPHER B. FIELD, Carnegie Institute of Washington, Stanford, California
JOHN GERHART, University of California, Berkeley
J. PAUL GILMAN, Celera Genomics, Rockville, Maryland
DANIEL S. GREENBAUM, Health Effects Institute, Cambridge, Massachusetts
BRUCE D. HAMMOCK, University of California, Davis
ROGENE HENDERSON, Lovelace Respiratory Research Institute, Albuquerque, New Mexico
CAROL HENRY, American Chemistry Council, Arlington, Virginia
ROBERT HUGGETT, Michigan State University, East Lansing
JAMES F. KITCHELL, University of Wisconsin, Madison
DANIEL KREWSKI, University of Ottawa, Ottawa, Ontario
JAMES A. MACMAHON, Utah State University, Logan
CHARLES O'MELIA, Johns Hopkins University, Baltimore, Maryland
WILLEM F. PASSCHIER, Health Council of the Netherlands, The Hague
ANN POWERS, Pace University School of Law, White Plains, New York
KIRK SMITH, University of California, Berkeley
TERRY F. YOSIE, American Chemistry Council, Arlington, Virginia

Senior Staff

JAMES J. REISA, Director
DAVID J. POLICANSKY, Associate Director and Senior Program Director for Applied Ecology
RAYMOND A. WASSEL, Senior Program Director for Environmental Sciences and Engineering
KULBIR BAKSHI, Program Director for the Committee on Toxicology
ROBERTA M. WEDGE, Program Director for Risk Analysis
JOHN HOLMES, Senior Staff Officer

COMMISSION ON LIFE SCIENCES

MICHAEL T. CLEGG *(Chair)*, University of California, Riverside
PAUL BERG *(Vice Chair)*, Stanford University, Stanford, California
FREDERICK R. ANDERSON, Cadwalader, Wickersham & Taft, Washington, D.C.
JOANNA BURGER, Rutgers University, Piscataway, New Jersey
JAMES E. CLEAVER, University of California, San Francisco
DAVID S. EISENBERG, University of California, Los Angeles
NEAL L. FIRST, University of Wisconsin, Madison
DAVID J. GALAS, Keck Graduate Institute of Applied Life Science, Claremont, California
DAVID V. GOEDDEL, Tularik, Inc., South San Francisco, California
ARTURO GOMEZ-POMPA, University of California, Riverside
COREY S. GOODMAN, University of California, Berkeley
JON W. GORDON, Mount Sinai School of Medicine, New York
DAVID G. HOEL, Medical University of South Carolina, Charleston
BARBARA S. HULKA, University of North Carolina, Chapel Hill
CYNTHIA J. KENYON, University of California, San Francisco
BRUCE R. LEVIN, Emory University, Atlanta, Georgia
DAVID M. LIVINGSTON, Dana-Farber Cancer Institute, Boston, Massachusetts
DONALD R. MATTISON, March of Dimes, White Plains, New York
ELLIOT M. MEYEROWITZ, California Institute of Technology, Pasadena
ROBERT T. PAINE, University of Washington, Seattle
RONALD R. SEDEROFF, North Carolina State University, Raleigh
ROBERT R. SOKAL, State University of New York, Stony Brook
CHARLES F. STEVENS, The Salk Institute for Biological Studies, La Jolla, California
SHIRLEY M. TILGHMAN, Princeton University, Princeton, New Jersey
RAYMOND L. WHITE, DNA Sciences, Inc., Mountain View, California

Staff

WARREN R. MUIR, Executive Director
JACQUELINE K. PRINCE, Financial Officer
BARBARA B. SMITH, Administrative Associate
LAURA T. HOLLIDAY, Senior Program Assistant

OTHER REPORTS OF THE
BOARD ON ENVIRONMENTAL STUDIES AND TOXICOLOGY

A Risk-Management Strategy for PCB-Contaminated Sediments (2001)
Toxicological Effects of Methylmercury (2000)
Strengthening Science at the U.S. Environmental Protection Agency:
 Research-Management and Peer-Review Practices (2000)
Scientific Frontiers in Developmental Toxicology and Risk Assessment
 (2000)
Modeling Mobile-Source Emissions (2000)
Toxicological Risks of Selected Flame-Retardant Chemicals (2000)
Copper in Drinking Water (2000)
Ecological Indicators for the Nation (2000)
Waste Incineration and Public Health (1999)
Hormonally Active Agents in the Environment (1999)
Research Priorities for Airborne Particulate Matter: I. Immediate Priorities
 and a Long-Range Research Portfolio (1998); II. Evaluating
 Research Progress and Updating the Portfolio (1999)
Ozone-Forming Potential of Reformulated Gasoline (1999)
Risk-Based Waste Classification in California (1999)
Arsenic in Drinking Water (1999)
Brucellosis in the Greater Yellowstone Area (1998)
The National Research Council's Committee on Toxicology: The First 50
 Years (1997)
Toxicologic Assessment of the Army's Zinc Cadmium Sulfide Dispersion
 Tests (1997)
Carcinogens and Anticarcinogens in the Human Diet (1996)
Upstream: Salmon and Society in the Pacific Northwest (1996)
Science and the Endangered Species Act (1995)
Wetlands: Characteristics and Boundaries (1995)
Biologic Markers (5 reports, 1989-1995)
Review of EPA's Environmental Monitoring and Assessment Program (3
 reports, 1994-1995)
Science and Judgment in Risk Assessment (1994)
Ranking Hazardous Waste Sites for Remedial Action (1994)
Pesticides in the Diets of Infants and Children (1993)
Issues in Risk Assessment (1993)
Setting Priorities for Land Conservation (1993)
Protecting Visibility in National Parks and Wilderness Areas (1993)
Dolphins and the Tuna Industry (1992)
Hazardous Materials on the Public Lands (1992)
Science and the National Parks (1992)

Animals as Sentinels of Environmental Health Hazards (1991)
Assessment of the U.S. Outer Continental Shelf Environmental Studies
    Program, Volumes I-IV (1991-1993)
Human Exposure Assessment for Airborne Pollutants (1991)
Monitoring Human Tissues for Toxic Substances (1991)
Rethinking the Ozone Problem in Urban and Regional Air Pollution (1991)
Decline of the Sea Turtles (1990)

Copies of these reports may be ordered from
the National Academy Press
(800) 624-6242
(202) 334-3313
www.nap.edu

OTHER REPORTS OF THE COMMITTEE ON TOXICOLOGY

Evaluating Chemical and Other Agent Exposures for Reproductive and Developmental Toxicity (2001)
Acute Exposure Guideline Levels for Selected Airborne Contaminants, Volume 1 (2000)
Review of the U.S. Navy's Human Health Risk Assessment of the Naval Air Facility at Atsugi, Japan (2000)
Methods for Developing Spacecraft Water Exposure Guidelines (2000)
Review of the U.S. Navy Environmental Health Center's Health-Hazard Assessment Process (2000)
Review of the U.S. Navy's Exposure Standard for Manufactured Vitreous Fibers (2000)
Re-Evaluation of Drinking-Water Guidelines for Diisopropyl Methylphosphonate (2000)
Submarine Exposure Guidance Levels for Selected Hydrofluorocarbons: HFC-236fa, HFC-23, and HFC-404a (2000)
Review of the U.S. Army's Health Risk Assessments for Oral Exposure to Six Chemical-Warfare Agents (1999)
Toxicity of Military Smokes and Obscurants, Volume 1(1997), Volume 2 (1999), Volume 3 (1999)
Assessment of Exposure-Response Functions for Rocket-Emission Toxicants (1998)
Toxicity of Alternatives to Chlorofluorocarbons: HFC-134a and HCFC-123 (1996)
Permissible Exposure Levels for Selected Military Fuel Vapors (1996)
Spacecraft Maximum Allowable Concentrations for Selected Airborne Contaminants, Volume 1 (1994), Volume 2 (1996), Volume 3 (1996), Volume 4 (2000)

Preface

Extremely hazardous substances (EHSs)[1] can be released accidentally as a result of chemical spills, industrial explosions, fires, or accidents involving railroad cars and trucks transporting EHSs. The people in communities surrounding industrial facilities where EHSs are manufactured, used, or stored and in communities along the nation's railways and highways are potentially at risk of being exposed to airborne EHSs during accidental releases. Pursuant to the Superfund Amendments and Reauthorization Act of 1986, the U.S. Environmental Protection Agency (EPA) has identified approximately 400 EHSs on the basis of acute lethality data in rodents.

As part of its efforts to develop acute exposure guideline levels for EHSs, EPA and the Agency for Toxic Substances and Disease Registry (ATSDR) requested that the National Research Council (NRC) in 1991 develop guidelines for establishing such levels. In response to that request, the NRC published *Guidelines for Developing Community Emergency Exposure Levels for Hazardous Substances* in 1993.

Using the 1993 NRC guidelines report, the National Advisory Committee (NAC) on Acute Exposure Guideline Levels for Hazardous Substances—consisting of members from EPA, the Department of Defense (DOD), the Department of Energy (DOE), the Department of Transportation, other federal and state governments, the chemical industry, academia, and other organizations from the private sector—has developed acute exposure guideline levels (AEGLs) for approximately 80 EHSs.

---

[1] As defined pursuant to the Superfund Amendments and Reauthorization Act of 1986.

In 1998, EPA and DOD requested that the NRC independently review the AEGLs developed by NAC. In response to that request, the NRC organized within its Committee on Toxicology the Subcommittee on Acute Exposure Guideline Levels, which prepared this report. This report evaluates the NAC's Standing Operating Procedures (SOP) document for its scientific validity, completeness, and for conformance to the 1993 NRC guidelines report. The report will be useful to EPA, DOD, ATSDR, and other federal, state, and local agencies, and industry in developing toxicologic risk assessments for hazardous chemicals.

This report has been reviewed in draft form by individuals chosen for their diverse perspectives and technical expertise, in accordance with procedures approved by the NRC's Report Review Committee. The purpose of this independent review is to provide candid and critical comments that will assist the institution in making its published report as sound as possible and to ensure that the report meets institutional standards for objectivity, evidence, and responsiveness to the study charge. The review comments and draft manuscript remain confidential to protect the integrity of the deliberative process. We wish to thank the following individuals for their review of this report: Gary Carlson (Purdue University), Charles Feigley (University of South Carolina), and Ralph Kodell (National Center for Toxicological Research).

Although the reviewers listed above have provided many constructive comments and suggestions, they were not asked to endorse the conclusions or recommendations nor did they see the final draft of the report before its release. The review of this report was overseen by Mary Vore (University of Kentucky), appointed by the Commission on Life Sciences, who was responsible for making certain that an independent examination of this report was carried out in accordance with institutional procedures and that all review comments were carefully considered. Responsibility for the final content of this report rests entirely with the authoring committee and the institution.

The subcommittee gratefully acknowledges the valuable assistance provided by the following persons: Roger Garrett, Paul Tobin, and Ernest Falke (all from EPA); George Rusch (Honeywell, Inc.); Po Yung Lu, Sylvia Talmage, Robert Young, and Cheryl Bast (all from Oak Ridge National Laboratory), and Karl Rozman (University of Kansas Medical Center). Aida Neel was the project assistant. Ruth Crossgrove edited the report. We are grateful to James J. Reisa, director of the Board on Environmental Studies and Toxicology (BEST), and David Policansky, associate director of BEST, for their helpful comments. The subcommittee particularly acknowledges Kulbir Bakshi, project director for the subcommittee, for bringing the report to completion. Finally, we would like to thank all members of the subcommittee for their expertise and dedicated effort throughout the development of this report.

Daniel Krewski, *Chair*
Subcommittee on Acute Exposure
Guideline Levels

Bailus Walker, *Chair*
Committee on Toxicology

Contents

INTRODUCTION *1*

ROSTER OF THE NATIONAL ADVISORY COMMITTEE FOR ACUTE
  EXPOSURE GUIDELINE LEVELS (NAC/AEGL COMMITTEE) FOR
  HAZARDOUS SUBSTANCES *7*

APPENDIX: STANDING OPERATING PROCEDURES FOR DEVELOPING
  ACUTE EXPOSURE GUIDELINE LEVELS FOR HAZARDOUS
  CHEMICALS *15*

PREFACE *17*

1. OVERVIEW OF AEGL PROGRAM AND NAC/AEGL COMMITTEE   *19*
   History, *19*
   Purpose and Objectives of the AEGL Program and the NAC/AEGL
     Committee, *21*
   Committee Membership and Organizational Structure, *23*
   Selection of Chemicals for AEGL Development, *24*
   Scientific Credibility of AEGL Values, *25*
   The AEGL Development and Peer-Review Process, *26*
   Operation of the NAC/AEGL Committee, *28*
   Value of a Collaborative Effort in the AEGL Program, *30*
   Applications of the AEGL Values, *31*

2. **DERIVATION OF AEGL VALUES** 34
   2.1 Characterization of AEGLs, *34*
   2.2 Empirical Toxicologic Endpoints and Methods for Determining Exposure Concentrations Used to Derive AEGLs 1,2, and 3, *35*
   2.3 Guidelines and Criteria for the Search Strategy, Evaluation, Selection, and Documentation of Key Data and Supporting Data Used for the Derivation of AEGL Values, *46*
   2.4 Dosimetry Corrections from Animal to Human Exposures, *57*
   2.5 Guidelines and Criteria for Selection of Uncertainty Factors to Address the Variability Between Animals and Humans and Within the Human Population, *62*
   2.6 Guidelines and Criteria for Selection of Modifying Factors, *91*
   2.7 Guidelines and Criteria for Time Scaling, *92*
   2.8 Guidelines and Criteria for Addressing Short-Term Exposure Known and Suspect Carcinogens, *111*
   2.9 Guidelines and Criteria for Miscellaneous Procedures and Methods, *122*

3. **FORMAT AND CONTENT OF TECHNICAL SUPPORT DOCUMENTS** 124
   Editorial Conventions, *124*
   3.1 Format and Content of Technical Support Documents, *125*
   3.2 Graphic Description of Data, *131*

4. **CURRENT ADMINISTRATIVE PROCESSES AND PROCEDURES FOR THE DEVELOPMENT OF AEGL VALUES** 139
   4.1 Committee Membership and Organizational Structure, *140*
   4.2 The AEGL Development and Peer-Review Process, *141*
   4.3 Operation of the NAC/AEGL Committee, *144*
   4.4 Role of the Director of the AEGL Program, *146*
   4.5 Role of the Designated Federal Officer, *147*
   4.6 Role of the NAC/AEGL Committee Chair, *147*
   4.7 Classification of the Status of AEGL Values, *148*
   4.8 Function of AEGL Development Teams, *148*
   4.9 Role of NAC/AEGL Committee Members, *151*
   4.10 Role of the Organization That Drafts TSDs, *152*

5. **REFERENCES** *154*

APPENDIX A. PRIORITY LISTS OF CHEMICALS. *165*

APPENDIX B. DIAGRAM OF THE AEGL DEVELOPMENT PROCESS *173*

APPENDIX C. GLOSSARY OF ACRONYMS, ABBREVIATIONS, AND SYMBOLS *175*

APPENDIX D. EXAMPLE OF A TABLE OF CONTENTS OF A TECHNICAL SUPPORT DOCUMENT *179*

APPENDIX E. EXAMPLE OF A SUMMARY OF A TECHNICAL SUPPORT DOCUMENT *182*

APPENDIX F. EXAMPLE OF THE DERIVATION OF AEGL VALUES APPENDIX IN A TECHNICAL SUPPORT DOCUMENT *186*

APPENDIX G. EXAMPLE OF TIME-SCALING CALCULATIONS APPENDIX IN A TECHNICAL SUPPORT DOCUMENT *190*

APPENDIX H. EXAMPLE OF A CARCINOGENICITY ASSESSMENT APPENDIX IN A TECHNICAL SUPPORT DOCUMENT *194*

APPENDIX I. EXAMPLE OF THE AEGL DERIVATION SUMMARY APPENDIX IN A TECHNICAL SUPPORT DOCUMENT *196*

APPENDIX J. LIST OF EXTANT STANDARDS AND GUIDELINES IN A TECHNICAL SUPPORT DOCUMENT *201*

LIST OF TABLES

TABLE 2-1 Values of n from Ten Berge et al. (1986). *94*
TABLE 3-1 Grouping Data into Categories for Plotting *132*
TABLE A-1 Priority List of Chemicals *168*

LIST OF FIGURES

FIGURE 1-1 Hazard assessment *33*
FIGURE 2-1 Decision tree for the selection of key and supporting data *55*
FIGURE 2-2 Effects of varying n in the equation $C^n \times t = k$ ... *104*
FIGURE 3-1 Plot of categories of data *137*
FIGURE 4-1 The AEGL development process *142*

Standing Operating Procedures for Developing *Acute Exposure Guideline Levels* for Hazardous Chemicals

Introduction

In the Bhopal disaster of 1984, approximately 2,000 residents living near a chemical plant were killed and 20,000 more suffered irreversible damage to their eyes and lungs following accidental release of methyl isocyanate. The toll was particularly high because the community had little idea what chemicals were being used at the plant, how dangerous they might be, and what steps to take in case of emergency. This tragedy served to focus international attention on the need for governments to identify hazardous substances and to assist local communities in planning how to deal with emergency exposures.

In the United States, the Superfund Amendments and Reauthorization Act (SARA) of 1986 required the U.S. Environmental Protection Agency (EPA) to identify extremely hazardous substances (EHSs) and, in cooperation with the Federal Emergency Management Agency and the Department of Transportation, to assist Local Emergency Planning Committees (LEPCs) by providing guidance for conducting health-hazard assessments for the development of emergency-response plans for sites where EHSs are produced, stored, transported, or used. SARA also required the Agency for Toxic Substances and Disease Registry (ATSDR) to determine whether chemical substances identified at hazardous waste sites or in the environment present a public-health concern.

As a first step in assisting the LEPCs, EPA identified approximately 400 EHSs largely on the basis of their "immediately dangerous to life and health" (IDLH) values developed by the National Institute for Occupational Safety and Health in experimental animals. Although several public and private groups, such as the Occupational Safety and Health Administration and the

American Conference of Governmental Industrial Hygienists, have established exposure limits for some substances and some exposures (e.g., workplace or ambient air quality), these limits are not easily or directly translated into emergency exposure limits for exposures at high levels but of short duration, usually less than 1 hr, and only once in a lifetime for the general population, which includes infants, children, the elderly, and persons with diseases, such as asthma, heart disease, or lung disease.

The National Research Council (NRC) Committee on Toxicology (COT) has published many reports on emergency exposure guidance levels and spacecraft maximum allowable concentrations for chemicals used by the Department of Defense (DOD) and the National Aeronautics and Space Administration (NASA) (NRC 1968; 1972; 1984a,b,c,d; 1985a,b; 1986a,b; 1987; 1988, 1994, 1996a,b; 2000a,b). COT has also published guidelines for developing emergency exposure guidance levels for military personnel and for astronauts (NRC 1986b, 1992). Because of the experience of COT in recommending emergency exposure levels for short-term exposures, EPA and ATSDR in 1991 requested that COT develop criteria and methods for developing emergency exposure levels for EHSs for the general population. In response to that request, the NRC assigned this project to the COT Subcommittee on Guidelines for Developing Community Emergency Exposure Levels for Hazardous Substances. The report of that subcommittee, *Guidelines for Developing Community Emergency Exposure Levels for Hazardous Substances* (NRC 1993), provides step-by-step guidance for setting emergency exposure levels for EHSs. Guidance is given on what data are needed, what data are available, how to evaluate them, and how to present the results.

In November 1995, the National Advisory Committee for Acute Exposure Guideline Levels for Hazardous Substances (NAC[1]) was established to identify, review, and interpret relevant toxicologic and other scientific data and to develop acute exposure guideline levels (AEGLs) for high-priority, acutely toxic chemicals. The NRC's previous name for acute exposure levels—community emergency exposure levels (CEELs)—was replaced by the term AEGLs to reflect the broad application of these values to planning, response, and prevention in the community, the workplace, transportation, the military, and the remediation of Superfund sites.

AEGLs represent threshold exposure limits for the general public and are applicable to emergency exposures ranging from 10 min to 8 h. Three levels—AEGL-1, AEGL-2, and AEGL-3—are developed for each of five

---

[1]NAC is composed of members from EPA, DOD, many other federal and state agencies, industry, academia, and other organizations. The roster of NAC is shown on page 7.

exposure periods (10 min, 30 min, 1 h, 4 h, and 8 h) and are distinguished by varying degrees of severity of toxic effects.

The three AEGLs are defined as follows:

AEGL-1 is the airborne concentration (expressed as ppm (parts per million) or mg/m^3 (milligrams per cubic meter)) of a substance above which it is predicted that the general population, including susceptible individuals, could experience notable discomfort, irritation, or certain asymptomatic nonsensory effects. However, the effects are not disabling and are transient and reversible upon cessation of exposure.

AEGL-2 is the airborne concentration (expressed as ppm or mg/m^3) of a substance above which it is predicted that the general population, including susceptible individuals, could experience irreversible or other serious, long-lasting adverse health effects or an impaired ability to escape.

AEGL-3 is the airborne concentration (expressed as ppm or mg/m^3) of a substance above which it is predicted that the general population, including susceptible individuals, could experience life-threatening health effects or death.

Airborne concentrations below AEGL-1 represent exposure levels that can produce mild and progressively increasing but transient and nondisabling odor, taste, and sensory irritation or certain asymptomatic, nonsensory effects. With increasing airborne concentrations above each AEGL, there is a progressive increase in the likelihood of occurrence and the severity of effects described for each corresponding AEGL. Although the AEGL values represent threshold levels for the general public, including susceptible subpopulations, such as infants, children, the elderly, persons with asthma, and those with other illnesses, it is recognized that individuals, subject to unique or idiosyncratic responses, could experience the effects described at concentrations below the corresponding AEGL.

REVIEW OF THE NAC DOCUMENT
STANDING OPERATING PROCEDURES ON ACUTE EXPOSURE GUIDELINE LEVELS FOR HAZARDOUS SUBSTANCES

Before developing AEGLs for individual chemicals, the NAC developed the guidelines document *Standing Operating Procedures of the National*

Advisory Committee on Acute Exposure Guideline Levels for Hazardous Substances (referred to as the SOP manual), which documents the procedures, methods, criteria, and other guidelines used by NAC in the development of the AEGL values. The information contained in the SOP document is based on the guidance provided by the NRC in its guidelines report (NRC 1993). The SOP document contains further details and clarification of specific procedures, methods, criteria, and guidelines interpreted from the NRC report.

In 1998, EPA and DOD asked the NRC to review the NAC's SOP document and AEGL reports for their scientific validity, completeness, and conformance to the 1993 NRC guidelines report. The NRC assigned this project to the COT Subcommittee on Acute Exposure Guideline Levels. The subcommittee members were chosen for their expertise in toxicology, epidemiology, pharmacology, medicine, industrial hygiene, biostatistics, risk assessment, and risk communication. The subcommittee's review of the SOP document prepared by the NAC involved oral and written presentations to the subcommittee by the authors of the report. The subcommittee provided advice and recommendations for revisions to ensure scientific validity and consistency with the NRC (1993) guidelines report. The authors of the SOP document presented their revised report at subsequent meetings until the subcommittee was satisfied with the revisions. The subcommittee concludes that the revised SOP document presented in the Appendix of this report is scientifically valid, complete, and consistent with the 1993 NRC guidelines report.

REFERENCES

NRC (National Research Council). 1968. Atmospheric Contaminants in Spacecraft. Washington, DC: National Academy of Sciences.

NRC (National Research Council). 1972. Atmospheric Contaminants in Manned Spacecraft. Washington, DC: National Academy of Sciences.

NRC (National Research Council). 1984a. Emergency and Continuous Exposure Limits for Selected Airborne Contaminants, Vol. 1. Washington, DC: National Academy Press.

NRC (National Research Council). 1984b. Emergency and Continuous Exposure Limits for Selected Airborne Contaminants, Vol. 2. Washington, DC: National Academy Press.

NRC (National Research Council). 1984c. Emergency and Continuous Exposure Limits for Selected Airborne Contaminants, Vol. 3. Washington, DC: National Academy Press.

NRC (National Research Council). 1984d. Toxicity Testing: Strategies to Determine Needs and Priorities. Washington, DC: National Academy Press.

NRC (National Research Council). 1985a. Emergency and Continuous Exposure

Guidance Levels for Selected Airborne Contaminants, Vol. 4. Washington, DC: National Academy Press.

NRC (National Research Council). 1985b. Emergency and Continuous Exposure Guidance Levels for Selected Airborne Contaminants, Vol. 5. Washington, DC: National Academy Press.

NRC (National Research Council). 1986a. Emergency and Continuous Exposure Guidance Levels for Selected Airborne Contaminants, Vol. 6. Washington, DC: National Academy Press.

NRC (National Research Council). 1986b. Criteria and Methods for Preparing Emergency Exposure Guidance Level (EEGL), Short-Term Public Emergency Guidance Level (SPEGL), and Continuous Exposure Guidance level (CEGL) Documents. Washington, DC: National Academy Press.

NRC (National Research Council). 1987. Emergency and Continuous Exposure Guidance Levels for Selected Airborne Contaminants, Vol. 7. Washington, DC: National Academy Press.

NRC (National Research Council). 1988. Emergency and Continuous Exposure Guidance Levels for Selected Airborne Contaminants, Vol. 8. Washington, DC: National Academy Press.

NRC (National Research Council). 1992. Guidelines for Developing Spacecraft Maximum Allowable Concentrations for Space Station Contaminants. Washington, DC: National Academy Press.

NRC (National Research Council). 1993. Guidelines for Developing Community Emergency Exposure Levels for Hazardous Substances. Washington, DC: National Academy Press.

NRC (National Research Council). 1994. Spacecraft Maximum Allowable Concentrations for Selected Airborne Contaminants, Vol. 1. Washington, DC: National Academy Press.

NRC (National Research Council). 1996a. Spacecraft Maximum Allowable Concentrations for Selected Airborne Contaminants, Vol. 2. Washington, DC: National Academy Press.

NRC (National Research Council). 1996b. Spacecraft Maximum Allowable Concentrations for Selected Airborne Contaminants, Vol. 3. Washington, DC: National Academy Press.

NRC (National Research Council). 2000a. Spacecraft Maximum Allowable Concentrations for Selected Airborne Contaminants, Vol. 4. Washington, DC: National Academy Press.

NRC (National Research Council). 2000b. Acute Exposure Guideline Levels for Selected Airborne Chemicals, Vol. 1. Washington, DC: National Academy Press.

Roster

National Advisory Committee for Acute Exposure Guideline Levels (NAC/AEGL Committee) for Hazardous Substances

COMMITTEE MEMBERS

George Rusch
Chair, NAC/AEGL Committee
Honeywell International, Inc.
Morristown, NJ

Ernest Falke
Chair, SOP Workgroup
U.S. Environmental Protection Agency
Washington, DC

George Alexeeff
California EPA
Oakland, CA

Steven Barbee
Arch Chemicals, Inc.
Norwalk, CT

Lynn Beasley
U.S. Environmental Protection Agency
Washington, DC

David Belluck
Minnesota Department of Transportation
Oakdale, MN

Robert Benson
U.S. Environmental Protection Agency
Denver, CO

Jonathan Borak
American College of Occupational and Environmental Medicine
and Yale University
New Haven, CT

William Bress
Vermont Department of Health
Burlington, VT

George Cushmac
U.S. Department of Transportation
Washington, DC

Larry Gephart
ExxonMobil Biomedical Sciences, Inc.
Annandale, NJ

Doan Hanson
Brookhaven National Laboratory
Upton, NY

John P. Hinz
U.S. Air Force
Brooks Air Force Base, TX

James Holler
Agency for Toxic Substances and Disease Registry
Atlanta, GA

Thomas C. Hornshaw
Illinois Environmental Protection Agency
Springfield, IL

Nancy K. Kim
New York State Department of Health
Troy, NY

Loren Koller
Oregon State University
Corvallis, OR

Glenn Leach
U.S. Army
Aberdeen Proving Grounds, MD

Mark A. McClanahan
Centers for Disease Control & Prevention
Doraville, GA

John Morawetz
International Chemical Workers Union
Cincinnati, OH

Richard Niemeier
National Institute for Occupational Safety and Health
Cincinnati, OH

Marinelle Payton
Harvard Medical School
Boston, MA

Zarena Post
Texas Natural Resource Conservation Commission
Austin, TX

George Rodgers
American Association of Poison Control Centers
University of Louisville
Louisville, KY

Robert Snyder
Environmental and Occupational Health Sciences Institute
Piscataway, NJ

Thomas J. Sobotka
U.S. Food and Drug Administration
Laurel, MD

Kenneth Still
U.S. Navy
Wright-Patterson Air Force Base, OH

Judy Strickland
U.S. Environmental Protection Agency
Research Triangle Park, NC

Richard Thomas
International Center for Environmental Technology
McLean, VA

Thomas Tuccinardi
U.S. Department of Energy
Germantown, MD

Past Committee Members

Kyle Blackman
Federal Emergency Management Agency
Washington, DC

Luz Claudio
Mount Sinai Medical Center
New York, NY

Benjamin Jackson
Consultant
Washington, DC

William Pepelko
U.S. Environmental Protection Agency
Washington, DC

Michelle Schaper
Mine Safety and Health Administration
Arlington, VA

Patricia Talcott
University of Idaho
Moscow, ID

Participants from Cooperative Partner Countries

Peter Griem
Forschungs- und Beratungsinstitut Gefarhstoffe
Freiburg, Germany

Ursula Gundert-Remy
Federal Institute for Consumer Health Protection and Veterinary Medicine
Berlin, Germany

Fritz Kalberlah
Forschungs- und Beratungsinstitut Gefarhstoffe
Freiburg, Germany

Marc Ruijten
Medical Emergency Preparedness and Planning Office
Rotterdam, The Netherlands

Ursula Stephan
Head, German Toxicology Expert Group for AEGLs
Berlin, Germany

Marcel van Raaij
RIVM Center for Chemical Risk Assessment
Rotterdam, The Netherlands

Other Participants

Surender Ahir
Occupational Safety and Health Administration
Washington, DC

Annick Pichard
Institut National de l'Environment Industriel et des Risques
Paris, France

Oak Ridge National Laboratory Staff

Po-Yung Lu
Manager, AEGL Project
Oak Ridge National Laboratory
Oak Ridge, TN

Cheryl B. Bast
Oak Ridge National Laboratory
Oak Ridge, TN

Kowetha Davidson
Oak Ridge National Laboratory
Oak Ridge, TN

Carol S. Forsyth
Oak Ridge National Laboratory
Oak Ridge, TN

Sylvia Milanez
Oak Ridge National Laboratory
Oak Ridge, TN

Robert H. Ross
Oak Ridge National Laboratory
Oak Ridge, TN

Sylvia Talmage
Oak Ridge National Laboratory
Oak Ridge, TN

Claudia M. Troxel
Oak Ridge National Laboratory
Oak Ridge, TN

Annetta Watson
Oak Ridge National Laboratory
Oak Ridge, TN

Robert A. Young
Oak Ridge National Laboratory
Oak Ridge, TN

AEGL Program Senior Staff

Roger Garrett
Director, AEGL Program
U.S. Environmental Protection Agency
Washington, DC

Paul S. Tobin
Designated Federal Officer, AEGL Program
U.S. Environmental Protection Agency
Washington, DC

Letty Tahan
Senior Scientist
U.S. Environmental Protection Agency
Washington, DC

Appendix

Standing Operating Procedures for Developing Acute Exposure Guideline Levels for Hazardous Chemicals

Preface

The National Advisory Committee for Acute Exposure Guideline Levels for Hazardous Substances (NAC/AEGL Committee) was established to develop scientifically credible short-term exposure limits for approximately 400 to 500 acutely toxic substances. These short-term exposure limits, referred to as acute exposure guideline levels, or AEGLs, are essential for emergency planning, response, and prevention of accidental releases of chemical substances. Further, it is important that the values developed be scientifically credible so that effective planning, response, and prevention can be accomplished.

To ensure scientific credibility, six major elements have been integrated into the AEGL development process. These elements are (1) adherence to the National Research Council (NRC) report *Guidelines for Developing Community Emergency Exposure Levels for Hazardous Substances*,[1] with changes or additions described in this Standing Operating Procedures Manual (SOP manual); (2) consideration of other NRC guidelines for developing short-term exposure limits; (3) the use of scientifically acceptable processes and methodologies to ensure consistent and scientifically credible AEGL values; (4) a comprehensive search and review of relevant data and information from both published and unpublished sources; (5) the extensive evaluation of the data and the development of AEGLs by a committee of scientific and technical

---

[1]NRC (National Research Council). 1993. Guidelines for Developing Community Emergency Exposure Levels for Hazardous Substances. Washington, DC: National Academy Press.

experts from both the public and private sectors; and (6) a multi-tiered peer-review process culminating with final review and concurrence by the NRC.

With the recent participation of certain member-countries of the Organization for Economic Cooperation and Development (OECD), it is anticipated that the AEGL program will be expanded to include the international community. That should result in increased scientific and technical support, a broader scope of the review process, and an even greater assurance of scientifically credible AEGL values.

This SOP manual represents the NAC/AEGL Committee's SOP Workgroup documentation of the procedures, methodologies, criteria, and other guidelines used by the NAC/AEGL Committee in the development of the AEGL values. The information contained herein is based on the guidance provided by the NRC (1993) guidelines report. This manual contains additions and further details and clarification of specific procedures, methodologies, criteria, and guidelines interpreted from the NRC guidelines that have been determined by the NAC/AEGL Committee to be a necessary supplement to the NRC guidelines. Procedures and methodologies included in this manual have been reviewed by the NAC/AEGL Committee and numerous OECD member countries and have received a review and concurrence by the NRC. New or modified procedures and methodologies that are developed and adopted by the NAC/AEGL Committee are classified as "proposed." Such procedures and methodologies will be submitted from time to time to the NRC for review and concurrence. Upon concurrence by the NRC, they will be considered final and will serve as a supplement to the 1993 NRC guidelines and to this manual.

It is believed that adherence to a rigorous AEGL development process in general and the use of scientifically sound procedures and methodologies in particular will provide the most scientifically credible exposure limits that are reasonably possible to achieve. This document is considered a "living document" and the various procedures and methodologies, including those classified as "final," are subject to change as deemed necessary by the NAC/AEGL Committee and the NRC Subcommittee on Acute Exposure Guideline Levels. As new data become available and new scientific procedures and methodologies become accepted by a majority of the relevant scientific community, the NAC/AEGL Committee, and the NRC, they will be integrated into the AEGL development process and the SOP manual. With this approach, both the scientific credibility of the AEGL values and the reduction in risk to the general population will be ensured.

1. Overview of AEGL Program and NAC/AEGL Committee

HISTORY

The concerns of the U.S. Environmental Protection Agency (EPA), other U.S. federal agencies, state and local agencies, private industry, and other organizations in the private sector regarding short-term exposures due to chemical accidents became sharply focused following the accidental release of methyl isocyanate in Bhopal, India in December of 1984. In November 1985, as part of EPA's National Strategy for Toxic Air Pollutants, EPA developed the Chemical Emergency Preparedness Program. This voluntary program identified a list of more than 400 acutely toxic chemicals and provided this information, together with interim technical guidance, for the development of emergency response plans at the local community level. At that time, EPA adopted the National Institute for Occupational Safety and Health (NIOSH) immediately dangerous to life and health (IDLH) exposure values, or an approximation of those values in instances where IDLH values were not published, to serve as the initial airborne concentrations of concern for each chemical.

During this same period, the U.S. Chemical Manufacturers Association (CMA) (now known as American Chemistry Council) developed and implemented the Community Awareness and Emergency Response Program. This program encouraged chemical plant managers to assist community leaders in preparing for potential accidental releases of acutely toxic chemicals. The program was intended to provide local communities with information on existing chemicals and chemical processes, technical expertise to assist in emergency planning, notification, and response, as well as in training emergency-response personnel.

In October 1986, as part of the reauthorization of Superfund, Congress wrote into law an emergency planning program under the Superfund Amendments and Reauthorization Act (SARA Title III). Under this act, states were required to have emergency response plans for chemical accidents developed at the local community level. EPA subsequently adjusted the level-of-concern values to one-tenth of the IDLH values or their equivalents as an approach to improving the safety of the levels used for the general public. Since that time, the agency and other organizations, including private industry, have been interested in adopting more rigorous methodologies for determining values that would be deemed safe for the general public. During this period, the American Industrial Hygiene Association (AIHA) established a committee, the Emergency Response Planning Guidelines (ERPG) Committee to develop ERPGs and pioneered the concept of developing three different airborne concentrations for each chemical that would reflect the thresholds for important health-effect endpoints. That committee was later renamed the Emergency Response Planning (ERP) Committee. Although constrained by limited resources, the ERP Committee has developed 1-h exposure limits for more than 70 chemicals during the past 10 years.

At a workshop hosted by EPA in 1987, it was proposed by EPA that the ERP Committee and scientists from federal and state agencies as well as scientists and clinicians from academia and public interest groups pool their technical and financial resources and form a single committee comprising scientists from both the public sector and the private sector to develop AEGL values. EPA conceived the idea to formulate general guidance for developing short-term exposure limits and together with the Agency for Toxic Substances and Diseases Registry (ATSDR) subsequently requested that the National Research Council (NRC) develop guidance on the use of procedures and methodologies to establish emergency exposure guideline levels for the general public. The NRC convened the Subcommittee on Acute Exposure Guideline Levels within the Committee on Toxicology (COT) to address the project.

Since the 1940s, COT has developed emergency exposure levels for numerous chemicals of concern to the U.S. Department of Defense (DOD). These values are referred to as "emergency exposure guidance levels" (EEGLs). Although the EEGLs were developed for use for military personnel, the NRC also developed emergency exposure levels for the general public, termed "short-term public emergency guidance levels" (SPEGLs). On the basis of this extensive experience and the high level of scientific and technical expertise continually available to the NRC, this organization was considered the most qualified entity to develop guidance on the methodologies and procedures used to establish short-term exposure limits for acutely toxic chemicals.

The NRC guidance document *Guidelines for Developing Community Emergency Exposure Levels for Hazardous Substances* was published in 1993. The community emergency exposure levels (CEELs) and the acute exposure guideline levels (AEGLs) represent the identical short-term emergency exposure levels. The NRC CEELs has been replaced by AEGLs to convey the broad applications of these values for planning, response, and prevention in the community, the workplace, transportation, the military, and the remediation of Superfund sites. A discussion of how AEGLs might be used for emergency planning, response, and prevention appears later in this chapter.

The efforts to mobilize the federal and state agencies and organizations in the private sector to form the committee began shortly after the release of the NRC (1993a) report. In October 1995, the National Advisory Committee on Acute Exposure Guideline Levels for Hazardous Substances (NAC/AEGL Committee) was formally chartered and the charter filed with the U.S. Congress under the Federal Advisory Committee Act (FACA) with approval by the Office of Management and Budget (OMB) and concurrence by the General Services Administration (GSA). Due to EPA budgetary constraints, the first meeting of the NAC/AEGL Committee was not held until June 1996. This meeting represented the culmination of the efforts to solicit stakeholders, identify committee members, form the committee, obtain the technical support of the Oak Ridge National Laboratories (ORNL), and begin the development of the AEGL values.

PURPOSE AND OBJECTIVES OF THE AEGL PROGRAM AND THE NAC/AEGL COMMITTEE

The primary purpose of the AEGL program and the NAC/AEGL Committee is to develop guideline levels for once-in-a-lifetime, short-term exposures to airborne concentrations of acutely toxic, high-priority chemicals. AEGLs are needed for a wide range of applications—planning, response, and prevention. These applications may include EPA's SARA Title III Section 302-304 emergency planning program, the U.S. Clean Air Act Amendments (CAAA) Section 112(r) accident prevention program, and the remediation of Superfund sites program; the Department of Energy's (DOE's) environmental restoration, waste management, waste transport, and fixed facility programs; the Department of Transportation's (DOT's) emergency waste response program; the Department of Defense's (DOD's) environmental restoration, waste management, and fixed facility programs; ATSDR's health consultation and risk assessment programs; NIOSH's and the Occupational Safety and Health Administration's (OSHA's) regulations and guidelines for workplace exposure; state CAA Section 112(b) programs and other state programs; and private-sector programs, such as the AIHA-ERPG and the American Chemistry Council's Chemtrec programs.

A principal objective of the NAC/AEGL Committee is to develop scientifically credible, acute (short-term) once-in-a-lifetime exposure guideline levels within the constraints of data availability, resources, and time. That objective requires highly effective and efficient efforts in data gathering, data evaluation, and data summarization; fostering the participation of a large cross-section of the relevant scientific community; and the adoption of procedures and methodologies that facilitate consensus-building for AEGL values within the committee.

Another principal objective of the committee is to develop AEGL values for approximately 400 to 500 acutely hazardous substances within the next 10 years. Therefore, the near-term objective is to increase the level of production of AEGL development to approximately 40 to 50 chemicals per year without exceeding budgetary limitations or compromising the scientific credibility of the values developed.

Further, in addition to determining AEGL values for three health-effect endpoints, it is intended to derive exposure values for the general public that are applicable to emergency (accidental), once-in-a-lifetime exposure periods ranging from 10 min to 8 h. Therefore, exposure limits will be developed for a minimum of five exposure periods (10 min, 30 min, 1 h, 4 h, and 8 h). Each AEGL tier is distinguished by varying degrees of severity of toxic effects, as initially conceived by the ERP Committee and further defined in *Guidelines for Developing Community Emergency Exposure Levels for Hazardous Substances* (NRC 1993a) and further defined by the NAC/AEGL Committee. The AEGL-1, AEGL-2, and AEGL-3 definitions are presented in the Introduction and in Section 2.1 of this standing operating procedures (SOP) manual.

As stated in the NRC guidelines (NRC 1993a) and described in the AEGL definitions, these exposure limits are intended to protect most individuals in the general population, including those that might be particularly susceptible to the deleterious effects of the chemicals. However, as stated in the guidelines and the definitions, it is recognized that certain individuals, subject to unique and idiosyncratic responses, could experience effects at concentrations below the corresponding AEGLs.

An important objective of the NAC/AEGL Committee is the establishment and maintenance of a comprehensive SOP manual that adheres to the NRC guidelines (NRC 1993a) and supplements, clarifies, interprets, or defines those guidelines with regard to the specific use of certain procedures and methodologies, such as selection of no-observed-adverse-effect levels (NOAELs) or the lowest-observed-adverse-effect levels (LOAELs), and the use of uncertainty factors, modifying factors, interspecies and intraspecies extrapolation methods, time scaling, carcinogenic risk assessment, and other methods and procedures relevant to the development of AEGL values.

COMMITTEE MEMBERSHIP AND ORGANIZATIONAL STRUCTURE

The NAC/AEGL Committee comprises representatives of federal, state and local agencies and organizations in the private sector that derive programmatic or operational benefits from the AEGL values. The membership includes federal representatives from EPA, ATSDR, DOE, NIOSH, OSHA, DOT, DOD, the Centers for Disease Control and Prevention (CDC), the Food and Drug Administration (FDA), and the Federal Emergency Management Agency (FEMA). States providing committee representatives include New York, New Jersey, Texas, California, Minnesota, Illinois, Connecticut, and Vermont. Private companies with representatives include Honeywell, Inc., ExxonMobil, and Arch Chemicals, Inc. The American Industrial Hygiene Association (AIHA), the American College of Occupational and Environmental Medicine (ACOEM), the American Association of Poison Control Centers (AAPCC), and the American Federation of Labor and Congress of Industrial Organizations (AFL-CIO) are represented, as are other private-sector organizations. The committee also includes individuals from academia and a representative of environmental justice. A current list of the NAC/AEGL Committee members and their affiliations is shown on page 7. At present, the committee has 30 members.

Recently, the Organization of Economic Cooperation and Development (OECD) and various OECD member countries have expressed an interest in the AEGL program. Several OECD member countries, such as Germany and The Netherlands, have been participating in the committee's activities and actively pursuing formal membership on the NAC/AEGL Committee. It is envisioned that the committee and the AEGL program in general will progressively expand its scope and participation to include the international community.

The program director of the AEGL program has the overall responsibility for the entire AEGL program and the NAC/AEGL Committee and its activities. A designated federal officer (DFO) is responsible for all administrative matters related to the committee to ensure that it functions properly and efficiently. These individuals are not voting members of the committee. The NAC/AEGL committee chair is appointed by EPA and is selected from the committee members. In concert with the program director and the DFO, the chair coordinates the activities of the committee and also directs all formal meetings of the committee. From time to time, the members of the committee serve as chemical managers and chemical reviewers in a collaborative effort with assigned scientist-authors (noncommittee members) to develop AEGLs for a specific chemical. These groups of individuals are referred to as the AEGL development teams and their function is discussed in Section 4.8 of this manual.

SELECTION OF CHEMICALS FOR AEGL DEVELOPMENT

A master list of approximately 1,000 acutely toxic chemicals was initially compiled through the integration of individual priority lists of chemicals submitted by each U.S. federal agency. The master list was subsequently reviewed by individuals from certain state agencies and representatives from organizations in the private sector and modified as a result of comments and suggestions received. The various priority chemical lists were compiled separately by each federal agency on the basis of their individual assessments of the hazards, potential exposure, risk, and relevance of a chemical to their program needs. A list of approximately 400 chemicals representing the higher-priority chemicals was tentatively identified from the original master list. It was acknowledged that this list was subject to change based on the changing needs of the stakeholders.

On May 21, 1997, a list of 85 chemicals was published in the *Federal Register*. This list identified those chemicals from the list of approximately 400 chemicals considered to be of highest priority across all U.S. federal agencies and represented the selection of chemicals for AEGL development by the NAC/AEGL Committee for the first 2-3 years of the program. The committee has addressed these chemicals, and they are in the "draft," "proposed," "interim," or "final" stages of development. Certain chemicals did not have an adequate database for AEGL development and, consequently, are on hold pending decisions regarding further toxicity testing. The initial highest-priority list of 85 chemicals is shown in Appendix A.

A second "working list" of priority chemicals is being selected from (1) the original master list, (2) the intermediate list of approximately 400 chemicals (which is a subset of the master list), and (3) the list of new, high-priority candidate chemicals submitted by U.S. agencies and organizations and OECD member countries that are planning to participate in the AEGL program. Although working lists will be published in the *Federal Register* and elsewhere from time to time to indicate the NAC/AEGL Committee's agenda, the priority of chemicals addressed, and, hence, the "working list," is subject to modification if priorities of the NAC/AEGL Committee or individual stakeholder organizations, including international members, change during that period.

SCIENTIFIC CREDIBILITY OF AEGL VALUES

The scientific credibility of the AEGL values is based on adherence to the NRC guidelines (NRC 1993a) for the development of short-term exposure limits, the comprehensive nature of data collection and evaluation, the consistency of the methodologies and procedures used to develop the values, the potential of acute toxicity testing in cases of inadequate data, and the adoption of the most comprehensive peer-review process ever used to establish short-term exposure limits for acutely toxic chemicals.

The comprehensive data-gathering process involves literature searches for all relevant published data and the mobilization of all relevant unpublished data. Data and information from unpublished sources is obtained through individual companies in the private sector and the cooperation of trade associations. The completeness of the data searches is enhanced through the oversight and supplemental searches conducted by individual committee members and interested parties during the peer-review process.

Data evaluation and data selection are performed by scientists with expertise in toxicology and related disciplines from staff at the organization that drafts "technical support documents" and from assigned members of the NAC/AEGL Committee. Additionally, input on data evaluation and selection is provided by interested parties who participate in the open meetings of the committee or who formally comment on the *Federal Register* notices of proposed AEGL values.

The work of the NAC/AEGL Committee adheres to *Guidelines for Developing Community Emergency Exposure Levels for Hazardous Substances* (NRC 1993a). Since this guidance document represents a more general guidance for methodologies and procedures, the NAC/AEGL Committee interprets and develops greater detail related to the methodologies and procedures that it follows. These standing operating procedures (SOPs) are documented by the SOP Workgroup and represent a consensus or a two-thirds majority vote of the NAC/AEGL Committee. SOPs also represent concurrence of the NRC Subcommittee on Acute Exposure Guideline Levels (NRC/AEGL Subcommittee). Therefore, each step of the AEGL development process follows specific methodologies, criteria, or other guidelines to ensure consistent, scientifically sound values.

In those instances where adequate data are not available in the judgment of the NAC/AEGL Committee, AEGLs will not be developed. The AEGL program is committed to ensuring that AEGL values are derived from adequate data and information based on a consensus or a two-thirds majority vote of the NAC/AEGL Committee and concurrence of the NRC/AEGL Subcommittee.

To further assure the scientific credibility of the AEGL values and their supporting rationale, the most comprehensive peer-review process ever used in the development of short-term exposure limits has been established (see next section). This review process has been designed to effectively, yet efficiently, encourage and enable the participation of the scientific community and other interested parties from the public and private sectors in the development of AEGLs. Further, the review process utilizes the NRC/AEGL Subcommittee as the final scientific reviewer. Hence, the final judgment of scientifically credible values rests with the United States' National Academy of Sciences' National Research Council. A detailed summary of the AEGL development process is presented in the next section.

THE AEGL DEVELOPMENT AND PEER-REVIEW PROCESS

The process that has been established for the development of the AEGL values is the most comprehensive ever used for the determination of short-term exposure limits for acutely toxic chemicals. A summary of the overall process is presented in diagram form in Appendix B. The process consists of four basic stages in the development and status of the AEGLs, and they are identified according to the review level and concurrent status of the AEGL values. They include (1) draft AEGLs, (2) proposed AEGLs, (3) interim AEGLs, and (4) final AEGLs. The entire development process can be described by individually describing the four basic stages in the development of AEGL values.

Stage 1: Draft AEGLs

This first stage begins with a comprehensive search of the published scientific literature. Attempts are made to mobilize all relevant unpublished data through industry-trade associations and from individual companies in the private sector. A more detailed description of the published and unpublished sources of data and information utilized is provided in Section 2.3 of this document, which addresses search strategies. The data are evaluated by following the published NRC guidelines (NRC, 1993a) and this SOP manual, and selected data are used as the basis for the derivation of the AEGL values and the supporting scientific rationale. Data evaluation, data selection, and development of a technical support document (TSD) are all performed as a collaborative effort among the staff scientists at the organization drafting the TSDs, the chemical manager, and two chemical reviewers. This group is called the AEGL Development Team. Specific NAC/AEGL Committee members are assigned to a team for each chemical under review. Hence, a separate team comprising different committee members is formed for each chemical under review. The product of this effort is a TSD that contains draft AEGLs. The draft TSD is subsequently circulated to all other NAC/AEGL Committee members for review and comment prior to a formal meeting of the committee. Revisions to the initial TSD and the draft AEGLs are made up to the time of the NAC/AEGL Committee meeting scheduled for formal presentation and discussion of the AEGL values and the documents. At the committee meeting, the committee deliberates and, if a quorum is present, attempts to reach a consensus or a two-thirds majority vote to elevate the draft AEGLs to "proposed" status. A quorum of the NAC/AEGL Committee is defined as 51% or more of the total NAC/AEGL Committee membership. If agreement cannot be reached, the committee conveys its issues and concerns to the AEGL Development Team and further work is conducted by this group. After completion of additional work, the chemical is resubmitted for consideration at a future meeting. If a consensus or a two-thirds majority vote of the committee cannot be achieved because of inadequate data, no AEGL values will be developed until adequate data become available.

Stage 2: Proposed AEGLs

Once the NAC/AEGL Committee has reached a consensus or a two-thirds majority vote on the AEGL values and supporting rationale, they are referred to as "proposed" AEGLs and are published in the *Federal Register* for a 30-day review and comment period. Following publication, the committee reviews the public comments, addresses and resolves relevant issues, and seeks a consensus or a two-thirds majority vote of those present on the original or modified AEGL values and the accompanying scientific rationale.

Stage 3: Interim AEGLs

Following resolution of relevant issues raised through public review and comment and subsequent approval of the committee, the AEGL values are classified as "interim." The interim AEGL status represents the best efforts of the NAC/AEGL Committee to establish exposure limits, and the values are available for use as deemed appropriate on an interim basis by federal and state regulatory agencies and the private sector. The interim AEGLs, the supporting scientific rationale, and the TSD, are subsequently presented to the NRC/AEGL Subcommittee for its review and concurrence. If concurrence cannot be achieved, the NRC/AEGL Subcommittee will submit its issues and concerns to the NAC/AEGL Committee for further work and resolution.

Stage 4: Final AEGLs

When concurrence by the NRC/AEGL Subcommittee is achieved, the AEGL values are considered "final" and published by the NRC. Final AEGL values may be used on a permanent basis by all federal, state and local agencies, and private organizations. It is possible that new data will become available from time to time that challenges the scientific credibility of final AEGLs. If that occurs, the chemical will be resubmitted to the NAC/AEGL Committee and recycled through the review process.

OPERATION OF THE NAC/AEGL COMMITTEE

The NAC/AEGL Committee meets formally four times each year for 2½ days each. The meetings are scheduled for each quarter of the calendar year and are generally held in the months of March, June, September, and December. Because of overall cost considerations, the meetings are generally held in Washington, D.C. However, committee meetings may be held from time to time at other locations for justifiable reasons.

At least 15 days before the committee meetings, a notice of the meeting is published in the *Federal Register* together with a list of the chemicals and other matters to be addressed by the committee. The notice provides dates, times, and location of the meetings. The agenda is finalized and distributed to committee members approximately 1 week before the meeting. The agenda is also available to other interested parties at that time, upon request, through the DFO.

All NAC/AEGL Committee meetings are open to the public, and interested parties may schedule individual presentations of relevant data and information by contacting the DFO to establish a date and time. Relevant data and information from interested parties also may be provided to the committee through the DFO during the period of development of the draft AEGLs so that the data can be considered during the early stages of development. Data and information also may be submitted when the document is in the proposed and interim stages.

The NAC/AEGL Committee meetings are conducted by the chair, who is appointed by EPA in accordance with the Federal Advisory Committee Act (FACA). At the time of the meeting, the chair and the other committee members will have received the initial draft and one or more revisions of the TSD and the draft, proposed, or interim AEGL values for each chemical on the agenda. Reviews, comments, and revisions are continuous up to the time of the meeting, and committee members are expected to be familiar with the draft, proposed, or interim AEGLs, the supporting rationale, and other data and information in each TSD, and to participate in the resolution of residual issues at the meeting. Procedures for the AEGL Development Teams and the other committee members regarding work on proposed or interim AEGLs are similar to those for draft AEGLs.

All decisions of the NAC/AEGL Committee related to the development of draft, proposed, interim, and final AEGLs and their supporting rationale are made by consensus or a two-thirds majority of a quorum of the committee members present.

The highlights of each meeting are recorded by the scientists who draft the TSDs, and written minutes are prepared, ratified, and maintained in the committee's permanent records. Deliberations of each meeting are also tape-recorded when possible and stored in the committee's permanent records by the DFO for future reference as necessary.

All proposed AEGL values and supporting scientific rationale are published in the *U.S. Federal Register*. Review and comment by interested parties and the general public are requested and encouraged. The committee's response to official comments on *Federal Register* notices on proposed AEGL values consists of an evaluation of the comments received, discussions and deliberations that take place at committee meetings regarding the elevation of AEGLs from proposed to interim status, and changes to the TSDs as deemed appropriate by the NAC/AEGL Committee. This information is reflected on the tapes and in the minutes of the meetings and will be maintained for future reference. As previously mentioned, the SOP Workgroup, established in March 1997, documents, summarizes, and evaluates the various procedures, methodologies, and guidelines used by the committee in the gathering and evaluation of scientific data and information and the development of the AEGL values. The SOP Workgroup performs a critical function by providing the committee with detailed information on the committee's interpretation of the NRC guidelines and the approaches the committee has taken in the derivation of each AEGL value for each chemical addressed. This documentation enables the NAC/AEGL Committee to continually document and assess the basis for its decision-making, ensure consistency with the NRC guidelines, and maintain the scientific credibility of the AEGL values and the accompanying scientific rationale. This ongoing effort is continually documented and is identified for further revisions in the SOP manual.

VALUE OF A COLLABORATIVE EFFORT IN THE AEGL PROGRAM

The value of a collaborative effort in the AEGL program is related primarily to the pooling of substantial resources of the various stakeholders and the direct or indirect involvement of a significant portion of the relevant scientific community from both the public and private sectors. These factors, in turn, promote greater productivity, efficiency, and cost effectiveness of such an effort and greatly enhance the scientific credibility of the AEGLs that are developed by the committee.

The formation of the NAC/AEGL Committee, which consists of approximately 30 to 35 members has provided an important forum for scientists, clinicians, and others to develop AEGLs and resolve related scientific issues. The composition of the committee represents a balanced cross-section of the relevant scientific disciplines and a balance of U.S. federal and state agencies, academia, the medical community, private industry, public interest groups, and other organizations in the private sector. This mutual participation of stakeholders, including the regulators and the regulated community, in the development of the AEGLs promotes the acceptance of the AEGLs by all parties involved. Additionally, the diverse composition of the committee represents the nucleus of a broad network of scientists, clinicians, and other technical personnel that fosters information and data exchange and the resolution of relevant scientific and technical issues well beyond the committee membership. This network also facilitates the identification of national and international experts with particular expertise that may provide important data, information, or insight on a specific chemical or scientific issue.

The collaborative effort also results in greater scientific credibility of the exposure values developed. The pooling of resources enables a comprehensive gathering and evaluation effort of both published and unpublished data and information. Collaboration provides a broad base of relevant scientific knowledge and expertise that is highly focused on the chemicals and issues addressed by the committee. This approach provides sufficient scientific and technical resources for the SOP Workgroup to document and evaluate procedures and methodologies that instill rigor and consistency into the process and the resultant AEGL values. The documentation of these procedures and methodologies is contained in this SOP manual. Finally, the collaborative effort has enabled the establishment of the most comprehensive peer-review process ever implemented for the development of short-term exposure limits.

The AEGL program has extended invitations to all OECD member countries to participate on the NAC/AEGL Committee and in program activities in general. To date, Germany and The Netherlands have commitments for participation. It is believed that expanding the scope of the AEGL program to include the international community will be of great benefit. Its participation will provide even greater resources, further broaden the base of scientific and technical expertise, provide new toxicologic data and insights, and foster the harmonization of emergency exposure limits at the international level.

In summary, the establishment of a collaborative effort, with its pooling of resources, represents the most productive, efficient, and cost-effective approach to the development of exposure guideline levels. Further, the effort results in the development of uniform values for a wide range of applications. This approach eliminates inconsistencies and confusion among individuals and organizations involved in emergency planning, response, and prevention of chemical accidents. In global terms, the NAC/AEGL Committee represents an approach to unifying the international community in the development and use of chemical emergency exposure limits. In the interests of multinational companies seeking uniform operating parameters and federal agencies having mandates to achieve international harmonization of standards and guidelines, the participation of the international community in the AEGL program represents an important goal of the AEGL program.

APPLICATIONS OF THE AEGL VALUES

It is anticipated that the AEGL values will be used for regulatory and nonregulatory purposes by U.S. federal and state agencies and possibly the international community in conjunction with chemical emergency response, planning, and prevention programs. More specifically, the AEGL values will be used for conducting various risk assessments to aid in the development of emergency preparedness and prevention plans, as well as real-time emergency response actions, for accidental chemical releases at fixed facilities and from transport carriers. The AEGL values, which represent defined toxic endpoints, are used in conjunction with various chemical-release and dispersion models to determine geographical areas, or "vulnerable zones," associated with accidental or terrorist releases of chemical substances. By determining these geographical areas and the presence of human populations and facilities within those zones, the potential risks associated with accidental chemical releases can be estimated. For example, the release and dispersion models, which take into account the quantity and rate of release of the chemical, the volatility of the substance, the wind speed and wind stability at the time of the release, and the topographical characteristics in the area of the release, will define the geographical areas exposed and, quantitatively, the airborne concentration of the "plume" or the chemical cloud as it is dispersed. By comparing the projected airborne concentrations of the chemical substance in question with the exposed populations, human health risks associated with a chemical release can be estimated. Using these risk estimates, emergency-response personnel can make effective risk-management and risk-communication decisions to minimize the adverse impact of the release on human health. Figure 1-1 is a summary diagram that indicates the overall effects that are expected to occur above each of the three AEGL threshold tiers as well as the sensory and nonsensory or asymptomatic effects that are expected to occur below the AEGL-1 threshold. Figure 1-1 also indicates the expected increase in occurrence and severity of the various adverse health effects as the airborne concentration increases beyond each of the three AEGLs.

Because of the complex nature of chemical accidents, the populations at risk, the variable capabilities among emergency response units, and many other considerations related to a specific event, it is beyond the scope of this document to discuss or speculate on specific actions that should or could be taken at any point in time or at a given level of exposure to a specific chemical. However, emergency responders and planners know that various options are available, depending upon the circumstances, for reducing or even preventing the adverse impacts of chemical releases. In general, they include public notification and instruction, sheltering-in-place, evacuation procedures to enable or facilitate medical attention, or a combination of these options. Decisions on these options are important and are best left to local emergency planners and responders to be addressed on a case-by-case basis. Further, information regarding the applications of short-term exposure limits such as AEGLs may be obtained in *Technical Guidance for Hazards Analysis* (EPA 1987).

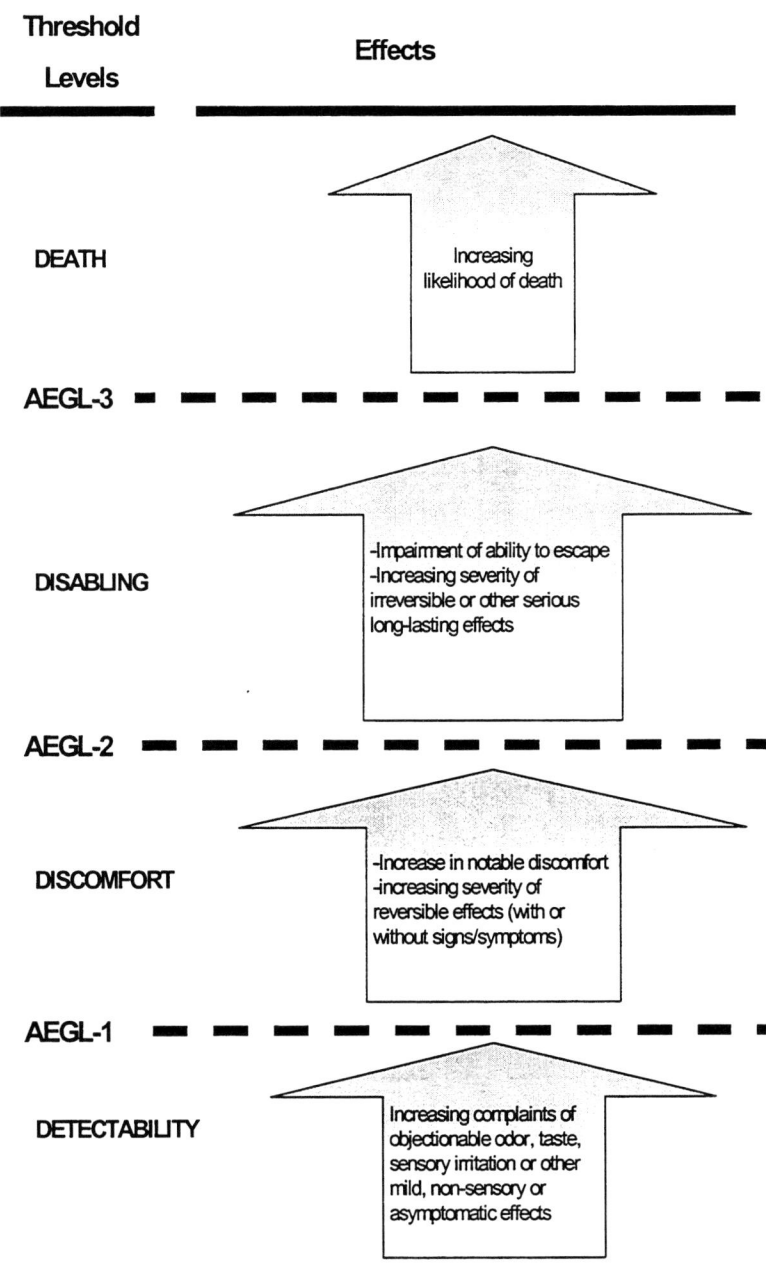

FIGURE 1-1 Hazard assessment.

2. Derivation of AEGL Values

2.1 CHARACTERIZATION OF AEGLs

Three tiers of AEGL severity levels represent short-term exposure values that are a threshold for specific biologic effects for the general public and are applicable to specified exposure durations of 10 min, 30 min, 1 h, 4 h, and 8 h. The values for these specified durations are maximum airborne concentrations above which there is an increasing likelihood of the adverse effects associated with the respective AEGL tiers. Therefore, to avoid the onset of these adverse effects, the values should not be exceeded during the specified exposure durations. Three tiers of AEGLs distinguished by varying degrees of severity of toxic effects are developed for each of the five exposure durations. Ten-minute AEGLs for the four chemicals included in the first publication of AEGLs by the National Research Council (NRC 2000b) will be developed at a future date.

Under the authority of the Federal Advisory Committee Act (FACA) P.L. 92-463 of 1972, the National Advisory Committee for Acute Exposure Guideline Levels for Hazardous Substances (NAC/AEGL Committee) has been established to identify, review, and interpret relevant toxicologic and other scientific data and develop AEGLs for high-priority acutely toxic chemicals.

AEGLs represent threshold exposure limits for the general public and are applicable to emergency exposure periods ranging from 10 min to 8 h. AEGL-2 and AEGL-3, and AEGL-1 values as appropriate, will be developed for each of five exposure periods (10 and 30 min, 1 h, 4 h, and 8 h) and will be distinguished by varying degrees of severity of toxic effects. It is believed that the recommended exposure levels are applicable to the general population including infants and children, and other individuals who may be susceptible. The three AEGLs have been defined as follows:

AEGL-1 is the airborne concentration (expressed as parts per million or milligrams per cubic meter (ppm or mg/m^3)) of a substance above which it is predicted that the general population, including susceptible individuals, could experience notable discomfort, irritation, or certain asymptomatic nonsensory effects. However, the effects are not disabling and are transient and reversible upon cessation of exposure.

AEGL-2 is the airborne concentration (expressed as ppm or mg/m^3) of a substance above which it is predicted that the general population, including susceptible individuals, could experience irreversible or other serious, long-lasting adverse health effects or an impaired ability to escape.

AEGL-3 is the airborne concentration (expressed as ppm or mg/m^3) of a substance above which it is predicted that the general population, including susceptible individuals, could experience life-threatening health effects or death.

Airborne concentrations below the AEGL-1 represent exposure levels that can produce mild and progressively increasing but transient and nondisabling odor, taste, and sensory irritation or certain asymptomatic, nonsensory effects. With increasing airborne concentrations above each AEGL, there is a progressive increase in the likelihood of occurrence and the severity of effects described for each corresponding AEGL. Although the AEGL values represent threshold levels for the general public, including susceptible subpopulations, such as infants, children, the elderly, persons with asthma, and those with other illnesses, it is recognized that individuals, subject to unique or idiosyncratic responses, could experience the effects described at concentrations below the corresponding AEGL.

2.2 EMPIRICAL TOXICOLOGIC ENDPOINTS AND METHODS FOR DETERMINING EXPOSURE CONCENTRATIONS USED TO DERIVE AEGLs 1, 2, AND 3

The selection of the biologic endpoints that serve as the thresholds for each of the AEGL severity levels are based on the definitions for the community emergency exposure levels (CEELs) that were published in the NRC guidelines for developing short-term exposure limits (NRC 1993a). The AEGLs address the same defined population as the NRC CEELs. The NRC definitions of the three CEEL tiers have been modified slightly by the NAC/AEGL Committee to improve only the clarity of description of the threshold levels. Hence, the defined threshold levels for CEELs and AEGLs are the same.

The NRC guidelines describe CEELs as exposure limits applicable to emergency exposures to hazardous substances (NRC 1993a). The NRC guidance states the CEELs must be set low enough to protect most of the population that may be exposed, including those with increased susceptibilities, such as children, pregnant women, persons with asthma, and persons with other specific illnesses (NRC 1993a). The NRC definition of CEELs for each of the three tiers of adverse health effects states that the adverse effects for each CEEL tier are not likely to occur below that level for a specified exposure duration but are increasingly likely to occur at concentrations above that level in a general population, including susceptible individuals. For this reason, the NRC also refers to the CEELs as threshold levels (NRC 1993a).

Because the data and methodologies used to derive AEGLs or any other short-term exposure limits are not sufficiently precise to make a distinction between a ceiling value and a threshold value, no distinction has been made with respect to AEGL values. No fine line can be drawn to precisely differentiate between a ceiling level, which represents the highest exposure concentration for which an effect is unlikely to occur, and a threshold level, which represents the lowest exposure concentration for the likelihood of onset of a given set of effects. Hence, AEGLs are not true effect levels. Rather, they are considered threshold levels that represent an estimated point of transition and reflect the best efforts to establish quantitatively a demarcation between one defined set of symptoms or adverse effects and another defined set of symptoms or adverse effects. Therefore, in the development of AEGLs, the NAC/AEGL Committee selects the highest exposure level from animal or human data where the effects used to define a given AEGL tier are not observed.

2.2.1 Selection of the Highest Exposure Level at Which the Effects That Define an AEGL Are Not Observed

Traditionally, when setting acceptable (typically considered "safe") levels of exposure, the risk assessor will select the highest experimental exposure that does not cause an adverse effect (no-observed-adverse-effect level (NOAEL)) in an experiment that demonstrated a graded exposure response from no effect to adverse effects. In standard risk-assessment practice (NRC 1993a), the exposure level identified as the NOAEL would then be divided by appropriate uncertainty factors and modifying factors to derive an acceptable exposure level for humans. However, there are a number of limitations in this method. It does not consider the number of animals used in the experiment and the associated statistical uncertainty around the experimental exposure level chosen. It does not consider the slope of the exposure-response relationship and subjects the risk assessor to using the possibly arbitrarily selected exposure levels chosen in the face of an unknown exposure-response relationship. Under some conditions, especially when only a small number of animals are exposed per exposure, the NOAEL could be a level associated with significant adverse health effects (Leisenring and Ryan 1992). In recent years, Crump and Howe (1984), Barnes et al. (1995), the U.S. Environmental Protection Agency (EPA 1995a), Faustman et al. (1994), Gaylor et al. (1998, 1999), and Fowles et al. (1999) addressed these problems by using the concept of analyzing all the data to estimate statistically a benchmark concentration (BMC). The BMC is a statistical estimate of an exposure that will cause a specified incidence of a defined adverse health effect. The BMC is commonly defined as the 95% lower confidence limit (LCL) on the exposure level causing a specified level of response (typically 1% to 10%). This exposure level is intended to replace the NOAEL and is used like the NOAEL when setting acceptable exposure levels.

The BMC method has a number of advantages over the traditional NOAEL approach. The BMC is derived from a statistical analysis of the exposure-response relationship and is not subject to investigator selection of exposure levels. It is a reflection of the exposure-response curve. Although the number of animals used in a study will affect the NOAEL and BMC estimates, the BMC, when compared with the maximum likelihood estimate (MLE), will explicitly reflect the variability in the study and the uncertainty around the number of subjects. The greater the variability and uncertainty, the greater the difference between the BMC and the MLE. The BMC calculation allows for the statistical estimation of a BMC in the absence of an empirical NOAEL.

The data most relevant to the development of AEGL-3 values and most amenable to a BMC analysis are inhalation LC_{50} (lethal concentration for 50% of the animals) data. Fowles et al. (1999) analyzed 120 inhalation animal lethality data sets by using the BMC method. The analysis provides the basis for the application of the BMC approach used by the NAC/AEGL Committee in the development of AEGL values. BMCs (95% LCL) and MLEs were developed for the 1%, 5%, and 10% response levels using log probit and Weibull models. Species tested included rats, mice, guinea pigs, hamsters, rabbits, and dogs. Exposure times ranged from 5 min to 8 h. Each data set consisted of at least four data points. The BMC and MLE values were compared with the empirical NOAEL (highest exposure that did not cause death in the experiment) and LOAEL (lowest exposure that killed at least one animal). The curve generated by the statistical models was subjected to a chi-squared goodness of fit test ($p > 0.05$). For statistical and data presentation reasons, 100 studies were analyzed with the probit analysis and 93 with the Weibull model. Most of the studies reported NOAELs (81/100 considered for the probit analysis and 74/93 considered for the Weibull analysis).

The BMCs were generally lower than the NOAELs when analyzed with either statistical estimate. The mean NOAEL/BMC ratios for the 1%, 5%, and 10% response were 1.60, 1.16, and 0.99, respectively, when using a probit analysis, and 3.59, 1.59, and 1.17, respectively, when using the Weibull analysis. It is interesting to note that comparable means from a Weibull analysis of developmental toxicity data were considerably greater, the developmental toxicity means of the NOAEL/BMC ratios were 29, 5.9, and 2.9 (Allen et al. 1994).

The proportion of times that the NOAEL exceeded the BMC for the 1%, 5%, and 10% response was 89%, 65%, and 42%, respectively, for the probit analysis and 95%, 80%, and 54%, respectively, for the Weibull analysis. In all cases, the LOAEL/BMC ratio exceeded 1 for the probit and Weibull analysis of the 1% and 5% response but not always for the 10% response (99%). For this reason, the BMC_{10} may be too high a response rate to use to predict a NOAEL. In contrast, the corresponding 1% and 5% response ratios were always greater than 1.

The ratios of the MLE/BMC were not great, ranging from a mean of 1.39 for a probit analysis of the 10% response to 3.02 for a Weibull analysis of the 10% response. It is important to note that when using the probit analysis, the LOAEL/MLE ratios were equal to or greater than 1 in 99%, 94%, and 71% of the cases for the 1%, 5%, and 10% responses, respectively. The MLE would probably be protective at the 1% response level but not for the 5% and 10% response levels. Similar numbers of 99%, 97%, and 76% were observed for the Weibull analysis.

The BMC approach can provide a more refined assessment of the prediction of the empirical NOAEL. It must be emphasized that even the empirical NOAEL may represent a response level that is not detected. When 5 to 10 animals are used in an experiment, a 10-20% response can be missed (Leisenring and Ryan 1992) and even a BMC_{10} is similar to a LOAEL with dichotomized data (Gaylor 1996). It is expected that the BMC is less than the empirical LOAEL. In the Fowles et al. (1999) analysis of the data, the BMC_{05} and BMC_{01} values were always below the empirical LOAEL for the studies analyzed. The probit analysis of the data by Fowles et al. (1999) provided a better fit with the data as measured by the "chi-squared goodness-of-fit test, mean width of confidence intervals, and number of data sets amenable to analysis by the model."

It is interesting to note that the BMC_{05} is very close to the MLE_{01} in the Fowles et al. (1999) evaluation of inhalation acute toxicity data. Through 1999, the NAC/AEGL Committee has used the MLE_{01} to estimate the highest exposure at which lethality is not likely to be observed in a typical acute exposure study. Given the analysis by Fowles et al. (1999) and for the above reasons, the NAC/AEGL Committee will generally use the BMC_{05} (lower 95% confidence limit (LCL) of the exposure required to produce a 5% response to exposure to chemicals) in the future for this estimate, although the MLE_{01} will also be calculated and considered. This approach incorporates the uncertainties due to the number of animals used in an experiment and the experimental variability observed; it utilizes all the data and the slope of the exposure-response curve and provides for a reasonable estimate of a predicted experimental NOAEL. In all cases, the MLE and BMC at specific response levels will be considered when setting AEGL values. Statistical models in addition to the log-probit will also be considered. Since goodness-of-fit tests consider an average fit, they may not be valid predictors of the fit in the low-exposure region of interest. In this case, the output of the different models will be plotted and compared visually with the experimental data in selection of the most appropriate model.

It should be emphasized that these methods will generally be considered for an acute lethal endpoint. Their use to set AEGL-1 and AEGL-2 values will be considered on a chemical-by-chemical basis. Different endpoints may require the use of different data sets in different or the same species, a different benchmark dose approach, or identification of a different response level. These factors will be considered for specific chemicals and toxicologic endpoints.

The preferred approach will be to use the BMC approach to identify the highest exposure at which the toxicologic effects used to define an AEGL tier were not observed. If the data are insufficient for a meaningful statistical analysis to use that approach, then the level will be determined empirically from experimental data.

2.2.2 Selection of Health-Effect Endpoints for AEGL-1, AEGL-2, and AEGL-3

In addition to the working definitions of the three AEGL tiers, this section includes a summary of the specific biologic endpoints used to establish the AEGL values for individual chemicals. Also included are general principles for selection of AEGL health-effect endpoints that have been derived from the committee's selections on a chemical-by-chemical basis. Since ideal data sets for certain chemicals are not available, extrapolation methods and the committee's scientific judgment are often used to establish threshold values. In the absence of adequate data, the committee can decide that no AEGL value be established. The basis for this decision is the failure to achieve a minimum two-thirds majority of a quorum of the committee that is in favor of establishing a value or a formal decision by two-thirds of the committee not to establish a value.

Under ideal circumstances, the specific health effects would be identified that determine each of the AEGLs. A search of the published literature for data on the chemical would be performed, and AEGLs would be generated from those data. However, data relating exposure and effect do not always follow an ideal paradigm and may lead to apparent inconsistencies in the use of endpoints to set AEGLs. The general principles laid down in the NRC (1993a) guidance for evaluating data and selecting appropriate health effects, combined with professional judgment, are used to establish AEGLs. From the evaluations of the first four chemicals in the subcommittee's first full report (NRC 2000b) and experience with data sets on chemicals currently under review, the following refinements to the NRC guidelines have been adopted by the NAC/AEGL Committee to set AEGLs.

For reasons discussed earlier in this chapter, the NAC/AEGL Committee generally selects the highest experimental concentration that does not elicit the symptoms or effects defined by the AEGL tier in question. This concentration represents the starting point for AEGL development. In instances in which appropriate data are available, the BMC method may be considered and used to select the AEGL endpoints.

2.2.2.1 AEGL-1 Endpoints

The NRC (1993a) guidelines discuss the definition of the CEEL-1 (AEGL-1) endpoint on pages 10, 12, and 21 of that report. Above the AEGL-1 value, discomfort becomes increasingly likely. Below the AEGL-1 value (detectability), "Exposure insufficient to cause discomfort or adverse health effects may be perceived nevertheless by means of smell, taste, or sensations (mild sensory irritation) that are not uncomfortable. The awareness of exposure may lead to anxiety and complaints and constitutes what is termed here detectability." (NRC 1993a, p. 21).

Thus, below the AEGL-1 values, there may be specific effects, such as the perception of a disagreeable odor, taste, or other sensations (mild sensory irritation). In some people, that exposure level could result in mild lacrimation or coughing. Since there is a continuum in which it is difficult to judge the appearance of "discomfort" in animal studies and human experiences, the

DERIVATION OF AEGL VALUES 41

NAC/AEGL Committee has used its best judgment on a case-by-case basis to arrive at appropriate and reasonable AEGL-1 values.

One additional factor to consider is that the three tiers of AEGL values "provide much more information than a single value because the series indicates the slope of the dose-response curve" (NRC 1993a). If an accident occurs and people smell or otherwise "detect" a chemical, the extent of the concentration range between AEGL-1 and AEGL-2 values provides useful information and insight into the estimated margin of safety between a level of detection or mild sensory irritation (AEGL-1) and a level that may impair escape or lead to a serious long-term or irreversible health effect (AEGL-2). In cases in which the biologic criteria for the AEGL-1 value would be close to, or exceed, the AEGL-2 value, the conclusion is reached that it is "not recommended" (NR) to develop AEGL-1 values. In these cases, "detectability" by itself would indicate that a serious situation exists. In instances in which the AEGL-1 value approaches or exceeds the AEGL-2 value, it may erroneously be believed that people experiencing mild irritation are not at risk when in fact they have already been exposed to extremely hazardous or possibly lethal concentrations.

Since a comparison of the AEGL-1 and AEGL-2 values indicates the slope of the dose-response curve that may be of value in emergency response, planning, or prevention, the NAC/AEGL Committee also attempts to establish AEGL-1 endpoints for adverse effects that are asymptomatic or nonsensory. Examples of such effects include significant (measurable) levels of methemoglobin, elevated blood enzyme levels, or other biologic markers related to exposure to a specific chemical. By establishing an AEGL-1 value in these instances, important information on the toxicologic behavior of a specific chemical is available to emergency responders and planners.

The following criteria have been used by the NAC/AEGL Committee to select endpoints for use in setting the AEGL-1 values.

2.2.2.1.1 No Value Established—AEGL-1 Is Close to or Exceeds AEGL-2

1. State what aspects of the chemical toxicity profile make it inadvisable to recommend an AEGL-1 value.

For example, the AEGL-1 value is not established, because levels that are "detectable" are close to, or exceed, an AEGL-2 value. These materials have poor warning properties.

2.2.2.1.2 No Value Established—Insufficient Data

Insufficient data were available to establish AEGL-1.

2.2.2.1.3 Highest Experimental Exposure Without an AEGL-1 Effect

1. State the species, effect, concentration, and exposure time to cause the effect.
2. Describe the toxicologic endpoint of concern.

For example, the highest experimental exposure levels that did not cause (a) sensory irritation, (b) altered pulmonary function, and (c) narcosis in humans have been used to set AEGL-1 values.

2.2.2.1.4 Effect Level for a Response

1. State the species, effect, concentration, and exposure time to cause the effect.
2. Describe the toxicologic endpoint of concern.

For example, levels for odor detection in humans, mild sensory irritation, asymptomatic or nonsensory effects, such as methemoglobin formation (22%) and altered pulmonary function (transient changes in clinically insignificant pulmonary functions of a susceptible individual), have been used as AEGL-1 endpoints.

2.2.2.2 AEGL-2 Endpoints

NRC (1993a) discussed the CEEL-2 (AEGL-2) definition on pages 10, 12, and 21 of its report. The AEGL-2 has been defined as the threshold between reversible effects that cause discomfort and serious or irreversible health effects or effects that impair escape. Above the AEGL-2 value, there is an increasing likelihood that people may become disabled or are increasingly likely to experience serious or irreversible health effects. "The term disability is used here to indicate the situation where persons will require assistance or where the effects of exposure will be more severe or prolonged without assistance" (NRC 1993a, p. 21). In developing AEGL-2 values, the NAC/AEGL Committee estimates a NOAEL for serious or irreversible effects or effects that impair escape. It must be emphasized that reversible clinical toxicity may be observed below the AEGL-2 value. If minor reversible effects are seen at one level of exposure and disabling effects at a higher exposure, the former is used to set the AEGL-2 value. If the exposure associated with disabling effects cannot be determined from experimental data, then the highest level causing reversible effects and discomfort may be used to set the AEGL-2 value.

DERIVATION OF AEGL VALUES 43

The following criteria have been used by the NAC/AEGL Committee to select endpoints for use in setting the AEGL-2 values.

2.2.2.2.1 Highest Experimental Exposure Without an AEGL-2 Effect

1. State the species, effect, concentration, and exposure time to cause the effect.
2. Describe the toxicologic endpoint of concern.

The highest experimental exposure levels that did not cause decreased hematocrit, kidney pathology, behavioral changes, or lethality (effects observed at higher exposures were above the definition for AEGL-2) have been used as the basis for determining AEGL-2 values.

2.2.2.2.2 Effect Level for a Toxic Response That Was Not Incapacitating or Not Irreversible

1. State the species, effect, concentration, and exposure time to cause the effect.
2. Describe the toxicologic endpoint of concern.

For example, overt ocular and/or respiratory tract irritation, dyspnea, pulmonary function changes, provocation of asthma episodes, pathology (respiratory tract), mild narcosis, and methemoglobin formation (approximately 40%) have been used as a basis for AEGL-2 values.

2.2.2.2.3 A Fraction of the AEGL-3 Value

1. State the rationale for using a fraction of the AEGL-3.
2. State why the specific fraction chosen is scientifically justified.

In the absence of specific data used to determine an AEGL-2 value, one-third of the AEGL-3 value has been used to establish the AEGL-2 value. This approach can only be used if the data indicate a steep exposure-based relationship based on data for effects below the AEGL-2 value and lethal-effect value.

2.2.2.3 AEGL-3 Endpoints

NRC (1993a) discussed the CEEL-3 (AEGL-3) definition on pages 10, 12, and 21 of its report. The AEGL-3 tier has been defined as the threshold exposure level between serious long-lasting or irreversible effects or effects that impair escape and death or life-threatening effects. Above the AEGL-3, there is an increasing likelihood of death or life-threatening effects occurring. In determining AEGL-3 values, the NAC/AEGL Committee defined the highest exposure level that does not cause death or life-threatening effects. It must be emphasized that severe toxicity will be observed at levels exceeding the AEGL-3. In cases in which data to determine the highest exposure level that does not cause life-threatening effects are not available, levels that cause severe toxicity without producing death have been used.

The following criteria have been used by the NAC/AEGL Committee to date to select endpoints for use in setting the AEGL-3 values.

2.2.2.3.1 Highest Exposure Level That Does Not Cause Lethality— Experimentally Observed Threshold (AEGL-3 NOAEL)

1. State the species, effect, concentration, and exposure time to cause the effect.
2. Describe the toxicologic endpoint of concern.

When experimental lethality data are insufficient to determine statistically a BMC, the highest experimental exposure that did not cause lethality in an experiment in which death was observed was used to set the AEGL-3 value.

2.2.2.3.2 Highest Exposure Level That Does Not Cause Lethality— Estimated Lethality Threshold—One-Third of the LC_{50}

1. State the species, effect, concentration, and exposure time to cause the effect.
2. Describe the toxicologic endpoint of concern.
3. If an exposure that does not produce death is estimated by dividing an LC_{50} value by 3 (or some other divisor), then give the slope of the exposure response curve or enough data points to support the division by 3 (or some other divisor).

When experimental lethality data have been insufficient to determine statistically an LC_{01} value, but an LC_{50} value was determined and all exposure levels caused lethality, a fraction of the LC_{50} value is used to estimate the threshold for lethality. In all cases, the exposure-response curve was steep, and the LC_{50} value was divided by 3. Fowles et al. (1999) analyzed 120 published inhalation animal lethality data sets using the BMC method. Their analyses of inhalation toxicity experiments revealed that for many chemicals the ratio between the LC_{50} and the experimentally observed nonlethal level was on average a factor of approximately 2, the 90th percentile was 2.9, and the 95th percentile was 3.5. There was a range of ratios from 1.1 to 6.5.

2.2.2.3.3 Highest Exposure Level That Does Not Cause Lethality— Benchmark Exposure Calculation of the 5% and 1% Response

1. State the species, effect, concentration, and exposure time to cause the effect.
2. Describe the toxicologic endpoint of concern.
3. State the statistical method used to derive a BMC_{05} and the MLE_{01}.

When sufficient information is available, the preferred method through 1999 utilized probit analyses (Finney 1971; Litchfield and Wilcoxon 1948; and the Number Cruncher Statistical System - Version 5.5) to determine the LC_{01} and the MLE was used for the LC_{01} value.

In the future, the BMC_{05} and MLE_{01} for lethality will be determined, presented, and discussed. Results from the above models will be compared with the log probit EPA (2000) benchmark dose software (http://www.epa.gov/ncea/bmds.htm). In all cases, the MLE and BMC at specific response levels will be considered. Other statistical models such as the Weibull may also be considered. Since goodness-of fit-tests consider an average fit, they may not be valid predictors of the fit in the low-exposure region of interest. In this case, the output of the different models will be plotted and compared visually with the experimental data to determine the most appropriate model. The method that results in values consistent with the experimental data and the shape of the exposure-response curve will be selected for AEGL derivations.

Because of uncertainties that may be associated with extrapolations beyond the experimental data, the estimated values are compared with the empirical data. Estimated values that conflict with empirical data will generally not be used.

2.2.2.3.4 Effect Level for a Response

1. State the species, effect, concentration, and exposure time to cause the effect.
2. Describe the toxicologic endpoint of concern.

When data are insufficient to estimate the highest exposure that does not cause lethality, exposure levels that cause severe toxicity in the absence of lethality are used in the selection of exposure levels to set AEGL-3 values.

The endpoints of concern include decreased hematocrit, methemoglobin formation (70-80%), cardiac pathology, and severe respiratory pathology.

2.3 GUIDELINES AND CRITERIA FOR THE SEARCH STRATEGY, EVALUATION, SELECTION, AND DOCUMENTATION OF KEY DATA AND SUPPORTING DATA USED FOR THE DERIVATION OF AEGL VALUES

2.3.1 Search Strategy

The literature search strategy focuses on three general sources of information: (1) electronic databases, primarily peer-reviewed journals, and government databases; (2) published books and documents from the public and private sectors of the United States and foreign countries, including references on toxicology, regulatory initiatives, and general chemical information; (3) data from private industry or other private organizations. The search strategy also includes the use of search terms to enhance the relevance of the electronic databases identified and retrieved.

(1) Electronic Database Coverage

The following databases are searched:

TOXLINE database (1981-current) from the U.S. National Library of Medicine's TOXNET
> TOXLINE covers the toxicologic effects of chemicals, drugs, and physical agents on living systems. Among the areas covered are adverse drug reactions, carcinogenesis, mutagenesis, developmental and reproductive toxicology, environmental pollution, and food contamination.

TOXLINE 65 database (1965-1980)
> Subject coverage is identical to TOXLINE for time periods that precede that of TOXLINE.

HAZARDOUS SUBSTANCES DATA BANK (HSDB) (Current) from TOXNET
> HSDB is a comprehensive factual and numeric chemical profile. Each chemical profile is peer reviewed for completeness and accuracy to reflect what is known about the chemical.

PUBLIC MEDLINE (PUBMED)

PUBMED includes MEDLINE and PREMEDLINE. MEDLINE, the U.S. National Library of Medicine's premier bibliographic database, covers medicine, nursing, dentistry, veterinary medicine, health-care systems, and the preclinical sciences. The above-mentioned TOXLINE searches include MEDLINE citations. PREMEDLINE, also produced by NLM, provides citation and abstract information before full records are added to MEDLINE. For a short period of time, this information is only available in PUBMED.

REGISTRY OF TOXIC EFFECTS OF CHEMICAL SUBSTANCES (RTECS)

RTECS, compiled by NIOSH (U.S. National Institute for Occupational Safety and Health), is a comprehensive database of basic toxicity information and toxic-effects data on more than 100,000 chemicals.

U.S. NATIONAL TECHNICAL INFORMATION SERVICE (NTIS)

The NTIS database provides access to the results of U.S. government-sponsored research, development and engineering, plus analyses prepared by federal agencies, their contractors, or grantees. It is a means through which unclassified, publicly available, unlimited distribution reports are made available from such U.S. agencies as National Aeronautics and Space Administration (NASA), Department of Energy (DOE), Department of Housing and Urban Development (HUD), Department of Transportation (DOT), and some 600 other agencies. In addition, some state and local government agencies contribute their reports to the database. NTIS also provides access to the results of government-sponsored research and development from other countries.

U.S. INTEGRATED RISK INFORMATION SYSTEM (IRIS)

IRIS contains data from EPA in support of human health risk assessment, focusing on hazard identification and dose-response assessment for specific chemicals.

U.S. FEDERAL RESEARCH IN PROGRESS (FEDRIP)

FEDRIP provides access to information about ongoing U.S. government-funded research projects in the fields of physical sciences, engineering, and life sciences.

U.S. DEFENSE TECHNICAL INFORMATION CENTER (DTIC)

DTIC is the central U.S. Department of Defense facility for access to scientific and technical information. The DTIC database includes technical reports, independent research and development summaries, technology transfer information, and research and development descriptive summaries. The scope of the DTIC collection includes areas normally associated with defense research, such as military sciences, aeronautics, missile technology, and nuclear science. The collection also includes information on biology, chemistry, environmental sciences, and engineering.

ORNL (U.S. Oak Ridge National Laboratory) IN-HOUSE DATABASES

CHEMICAL UNIT RECORD ESTIMATES (CURE)
The CURE database contains selected information from the EPA Office of Health and Environmental Assessment documents and Carcinogen Risk Assessment Verification Effort (CRAVE) and Reference Dose (RfD) work groups. Although the groups are not currently active, this database is a valuable compilation of historic information.

TOXICOLOGY AND RISK ANALYSIS (TARA) DOCUMENT LIST
This database lists all types of documents written by TARA staff over the past 15 years. These range from toxicity summaries to journal articles. This list provides good references for chemicals that overlap the AEGL priority list.

(2) Published Books and Documents from the Public and Private Sectors

GENERAL REFERENCES FOR TOXICOLOGY AND CHEMICAL INFORMATION
ATSDR (U.S. Agency for Toxic Substances and Disease Registry) Toxicological Profiles.
Chemfinder, Chemical Searching and Information Integration by CambridgeSoft Corporation
Current Contents, Life Sciences edition
HEAST (Health Effects Assessment Summary Tables)
Kirk-Othmer Encyclopedia of Chemical Technology
IARC (International Agency for Research on Cancer) Monographs on the Evaluation of the Carcinogenic Risk of Chemicals to Humans
Low-Dose Extrapolation of Cancer Risks, S. Olin et al. (editors)
Merck Index

NTP (U.S. National Toxicology Program) Division of Toxicology Research and Testing, published reports.
Patty's Industrial Hygiene and Toxicology
Respiratory System, Monographs on the Pathology of Laboratory Animals, T.C. Jones et al. (editors)
Synthetic Organic Chemicals, U.S. International Trade Commission
Toxicology of the Nasal Passages, C.S. Barrow (editor)
U.S. Air Force Installation Restoration Program Toxicology Guide

GENERAL REFERENCES FOR REGULATORY INFORMATION AND STANDARDS
AIHA (American Industrial Hygiene Association) Emergency Response Planning Guidelines (ERPGs) and Workplace Exposure Level Guides (WEELs)
ACGIH (American Conference of Government and Industrial Hygienists) Threshold Limit Values for Chemical Substances and Physical Agents and Biological Exposure Indices
ACGIH Documentation of Threshold Limit Values
NAAQS (U.S. National Ambient Air Quality Standards)
NIOSH (U.S. National Institute for Occupational Safety and Health) Documentation of IDLH's (immediately dangerous to life and health)
NIOSH Pocket Guide to Chemical Hazards
NIOSH Recommendations for Occupational Safety and Health, Compendium of Policy Documents and Statements
OSHA (U.S. Occupational Safety and Health Administration) Limits for Air Contaminants
SMACs (Spacecraft Maximum Allowable Concentrations for Selected Airborne Contaminants), Committee on Toxicology, sponsored by the National Research Council
EPA Health Effects Documents

(3) Unpublished Data from Private Industry and Other Private Sector Organizations of All Nations

These data consist of reports and data that are not published in peer-reviewed scientific journals but are relevant to the development of AEGLs. Most often, the data represent acute toxicity data from controlled inhalation exposure studies available from private industry or other organizations in the private sector of all nations that may or may not be published in a peer-reviewed journal.

Search Terms

The U.S. Chemical Abstract Services (CAS) Registry number of the chemical is used as the first choice. Chemical nomenclature or common chemical names and synonyms are used if the CAS Registry number is unknown.

The CAS Registry number alone is used as the first step. If there are approximately 300 citations, then all are retrieved for review. If less than approximately 300 references are found, searches are conducted using chemical nomenclature and common chemical names in addition to the CAS number. Searches by chemical name or names also should be made if few data of high quality are found, irrespective of the number of citations found.

If more than 300 citations are found using any form of chemical identification, the references may be enriched in relevance and quality by adding any number of the following characterizations of the desired data to the search strategy:

- short-term
- threshold limit
- permissible exposure limit
- acute toxicity
- ocular terms
- inhalation terms
- dermal terms

If the number or quality of single-exposure toxicity studies found is not deemed to be adequate, multiple exposure studies may be considered but may not achieve a consensus of the NAC/AEGL Committee. If a consensus or two-third majority of the committee cannot agree on the adequacy of the data, the chemical may be placed on the list for future acute toxicity testing.

2.3.2 Evaluation, Selection, and Documentation of Key and Supporting Data

As a detailed interpretation and supplementation of the NRC (1993a) guidelines, the NAC/AEGL Committee has developed guidelines for evaluating the quality of studies to be used in the calculation of proposed AEGL values. The proposed evaluation and documentation procedure created by the NAC/AEGL Committee is intended to provide technical-support-document writers, reviewers, committee members, interested parties, and the public with

DERIVATION OF AEGL VALUES   51

a clear and consistent list of elements that must be considered in their evaluations. The proposed evaluation and documentation system will add technical validity and administrative credibility to the process by providing a transparent, logical, and consistent methodology for selecting key studies used to calculate an AEGL value. Additionally, the system will allow the proper selection of uncertainty factors and modifying factors in a consistent and logical manner. The process is designed to allow maximum flexibility in professional judgment while promoting scientific uniformity and consistency and providing a sound administrative foundation on which committee members can function.

Many toxicology studies used in the development of an AEGL were not designed to meet current regulatory guidelines and are not necessarily consistent in protocol or scientific methodology. As a result, these valuable investigations cannot be judged solely on the basis of currently accepted experimental design criteria for such studies. Current guidelines from EPA (1998) and the Organization for Economic Cooperation and Development (OECD 1993) are used as the basis for conducting future studies on behalf of the NAC/AEGL Committee, but lack of consistency of older studies requires evaluation and qualification of each data set for scientific validity within the context of AEGL documentation. A study can be valuable in the derivation of AEGLs without conforming completely to a standard of detailed methodology, data analysis, and the results reported. The aim of the subject procedure is to provide specific criteria in the selection and use of specific data sets for development of defensible values, yet retain the ability to use logical scientific thinking and competent professional judgment in the data-selection process. If a study or some portion of a study uses scientifically valid methods, contains adequate and reliable data, and presents defensible conclusions relevant to the AEGL process, it may be included in the technical support document (TSD) and used to support the AEGLs.

It is important to emphasize that only toxicity data obtained directly from a primary reference source are used as the basis for "key" toxicity studies from which the AEGL values are derived. Additionally, all supporting data and information important to the derivation of an AEGL value are obtained solely from the primary references. These data include those used to provide a "weight-of-evidence" rationale in support of the AEGL value derived. Secondary references may be used to provide data and information on commercial uses, production volumes, chemical and physical properties, and other nontoxicologic or epidemiologic information on a chemical. Secondary references also may be used to present background information on the toxicity of a chemical. Any other information not important or directly relevant to the actual derivation of the AEGL values may be used to provide supporting rationale for the AEGL values. Data and information from secondary references should not be included in data summary tables presented in the TSDs

The credibility of evaluation guidelines is enhanced when they are drawn from a widely accepted prescription for study protocol design. The guidelines for a study evaluation should be based on the scientific methodologies but not be so restrictive that it precludes competent professional judgment. Current Good Laboratory Practice (GLP) guidelines provide a basis for selection of a robust list of study elements that, in concert with the professional experience and judgment of the AEGL Development Team and NAC/AEGL Committee members in general, are used to qualify the data which support the AEGLs. Consequently the NAC/AEGL Committee has used the NRC guidelines (1993a), the OECD Guidelines for the Testing of Chemicals (OECD 1993), and the EPA Health Effects Test Guidelines (EPA 1998) as a basis for selection.

The NRC (1993a) guidance provides general insight on the use of toxicologic data for AEGL derivation from routes of exposure other than inhalation. The NRC (1993a) guidance states that the bioavailability and differences in the pharmacokinetics from different exposure routes of the chemical in question must be considered. Because of these complex biologic phenomena and the paucity of data to enable credible evaluation and consideration, the NAC/AEGL Committee to date has selected and used only inhalation toxicity data to derive AEGL values. Further, toxicity data on other exposure routes will not be included in discussions in the TSDs unless those data are considered important for the support of relevant pharmacokinetic or metabolism data or mechanisms of toxicity. In the absence of inhalation data to derive an AEGL value, the NAC/AEGL Committee may use toxicity data from other exposure routes if there are adequate data to perform scientifically credible route-to-route extrapolations. In the absence of acceptable data, the committee will refer the chemical for toxicity testing.

Each key and supporting study is evaluated using all listed "elements for evaluation" as guidance. A "key study" is defined as the human and/or animal study from which a toxicologic value is obtained for use in AEGL calculations. "Supporting studies" include those human and/or animal studies used to support the toxicologic findings and values obtained from the key study, and their use is consistent with the weight-of-evidence approach to scientific credibility. While all elements for evaluation listed below are considered when evaluating a study, only elements for evaluation from key and supporting studies that are relevant to the derivation of the AEGL values will be discussed in the TSD as they impact the derivation. In evaluating a study, a variety of endpoints are preferred. However, a study measuring, for example, only one endpoint may be selected for development of an AEGL if other studies have shown that other known inhalation toxicology endpoints are less sensitive, provided the data are considered reliable. The list of elements for evaluation is used for initial review of all studies evaluated for possible inclusion in the TSD in instances in which they are germane to the selection of studies.

The NAC/AEGL Committee is dependent upon existing clinical, epidemiologic, and case report studies published in the literature for data on humans. Many of these studies do not necessarily follow current guidelines on ethical standards that require effective, documented, informed consent from participating humans subjects. Further, recent studies that followed such guidelines may not include that fact in the publication. Although human data may be important in deriving AEGL values that protect the general public, utmost care must be exercised to ensure first of all that such data have been developed in accordance with ethical standards. No data on humans known to be obtained through force, coercion, misrepresentation, or any other such means will be used in the development of AEGLs. The NAC/AEGL Committee will use its best judgment to determine whether the human studies were ethically conducted and whether the human subjects were likely to have provided their informed consent. Additionally, human data from epidemiologic studies and chemical accidents may be used. However, in all instances described here, only human data, documents, and records will be used from sources that are publicly available or if the information is recorded by the investigator in such a manner that subjects cannot be identified directly or indirectly. These restrictions on the use of human data are consistent with the "Common Rule" published in the *Code of Federal Regulations* (Protection of Human Subjects, 40 CFR 26, 2000).

In addition to the discussion of the elements for evaluation in the individual studies section of the TSD, a section entitled "Data Adequacy and Research Needs" is included in the text of the TSD. A summary of the data-adequacy discussion is also included in the derivation summary tables in the appendix of the TSD and in the summary section of the TSD. The text of the TSD relates the studies used to derive or support the derivation of the AEGL values to the discussion of the adequacy of the available data. Brief summaries of this discussion are included in the summary and derivation summary tables. The data-adequacy section also presents and integrates the weight of evidence by considering all information as a whole for each AEGL developed. In addition to considering the elements for evaluation as relevant in the discussion, a number of other factors must be considered. They include repeatability of experiments between laboratories, consistency of data between experiments and laboratories, types and number of species tested, variability of results between species, and comparison of AEGL values with the valid human and animal data. Every data set is a unique, chemical-specific source of information that reflects the investigations conducted on the chemical and the properties of the chemical. This section reflects a "best professional judgment" approach in the evaluation of the data adequacy and future research needs.

Figure 2-1 contains a diagram of the decision process for the selection of key studies and supporting studies. A summary of the elements or criteria used to select key studies and supporting studies and to evaluate their adequacy in deriving AEGL values follows.

Elements for the Evaluation of Key and Supporting Data and Studies

1. Only toxicity data and information obtained directly from a primary reference source may be used as the basis for "key" toxicologic studies. All other studies important to the derivation of an AEGL value or that serve as a weight-of-evidence rationale are obtained from a primary source.
2. Secondary references may be used for nontoxicologic data, such as physical and chemical properties, production locations, quantities, and background information on the toxicity of a chemical, provided the information is not directly used in the derivation of the AEGL values.
3. Only human data from studies that meet the ethical standards discussed in the "Evaluation, Selection, and Documentation of Key and Supporting Data" section of this SOP manual will be used in the derivation of AEGL values.
4. The inhalation route of exposure is preferred. When the endpoint of concern is systemic toxicity and the hepatic first-pass effect is not significant, oral exposure may be considered. In the absence of scientifically sound inhalation data and with high confidence in a valid route-to-route extrapolation, routes of exposure other than inhalation will not be used for AEGL derivation.
5. Scientifically credible exposure concentration and exposure duration for the study are provided.
6. Analytical procedures are used to determine chamber concentration for inhalation exposure in controlled studies, and detailed, scientifically credible methods, procedures, and data are used to measure chemical concentration in epidemiologic or anecdotal cases (accidental chemical releases). For oral exposure, dose may be determined from the amount of test chemical placed into the subject.

DERIVATION OF AEGL VALUES

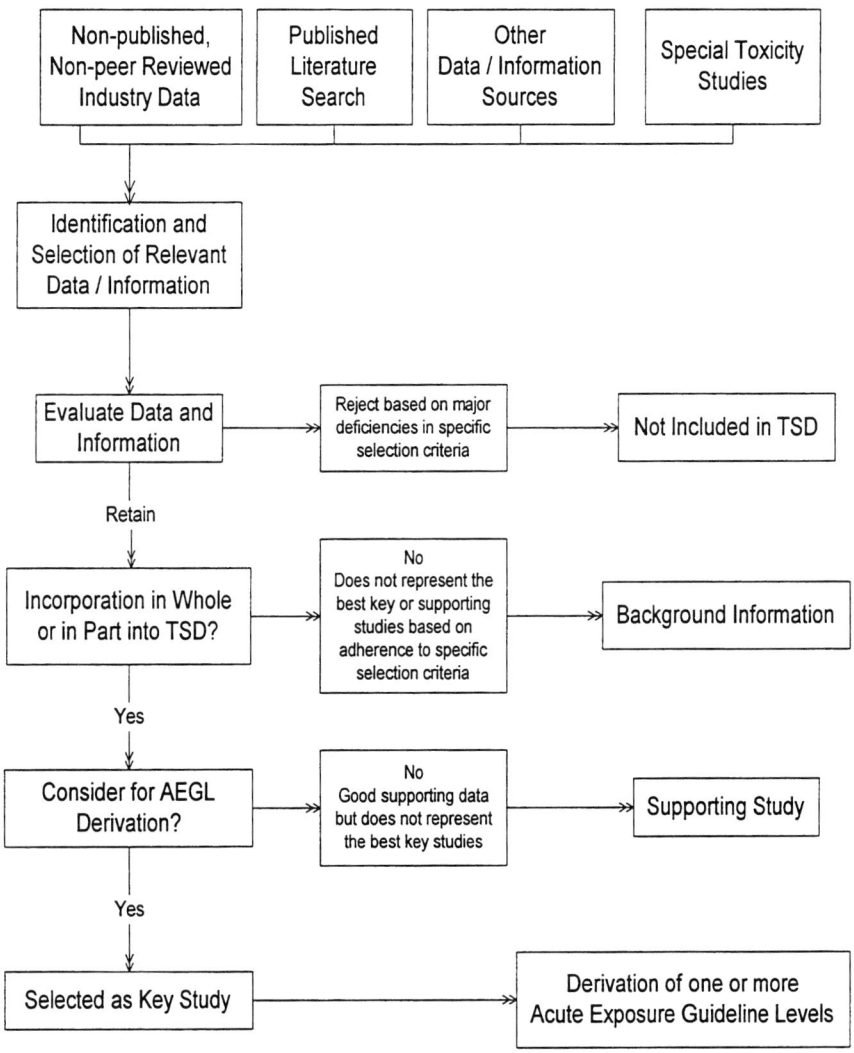

FIGURE 2-1 Decision tree for the selection of key and supporting studies.

7. The number of subjects is not rigid; e.g., a general rule uses 5-10 rodents/sex/group as a valid measure, but as few as 2-3 primates or dogs/sex/group may be used. The acceptable number of subjects per group is influenced by the relationship between the within-group variability and the degree of change that is considered to be detrimental. Smaller numbers

per group may be acceptable by increasing the number of treatment groups.
8. Humans are the most relevant species studied. Rats, mice, rabbits, guinea pigs, ferrets, dogs, or monkeys are acceptable. Other species require evaluation on a case-by-case basis. It is important to use a species for which there are historical control data and relevance to humans.
9. A concurrent control group is composed of the same species as that in the treatment groups. The control subjects should be housed and cared for in the same manner as exposed animals.
10. Concentration or dose selection establishes a clear dose-response relationship.
11. The observation period is variable based on the time of onset of the toxic effect. If it is rapid (minutes to 2-3 h) and associated with quick recovery, an observation period of 3-4 days may be sufficient. For effects that are slow in onset (2-3 days) and delayed in time, a minimum observation period of 14 days is recommended.
12. Signs and symptoms of toxicity are noted during and after exposure and reported separately by sex and concentration or dose.
13. For animal studies, body weights should be recorded throughout the study.
14. For repeated concentration or dose studies, the highest estimated or experimental (empirical) level of no effect is established for the specific AEGL endpoint of concern.
15. Toxicity data from routes of exposure other than inhalation generally will not be used as key or supporting data. Data from alternate routes are considered in the absence of inhalation data if sufficient data are available to perform a credible route-to-route extrapolation.
16. Number of concentrations or doses used are given.
17. If a NOAEL is selected or derived as the endpoint for an AEGL severity level of concern, identifying both the highest dose at which the effect is not seen and the lowest dose at which it is seen for each AEGL severity level strengthens the confidence in the study.
18. Time of death is recorded if applicable.
19. For animal studies, necropsy is conducted with at least gross examination results noted.
20. As available, data (e.g. histopathologic changes, clinical chemistry, and hematology) may reduce uncertainty.
21. Recovery group included in the study and data generated are sufficient to determine the degree of reversibility.
22. There is statistical treatment of data generated from study.
23. An evaluation of all relevant data should be performed and summarized in the TSDs to present an integrated weight-of-evidence picture for all information considered as a whole.

2.3.3 Elements for Discussion on Data Adequacy and Research Needs

The adequacy of the key and supporting data selected for AEGL derivation should be discussed in Section 8.3 of the TSD ("Data Adequacy and Research Needs"). Because of the different toxic endpoints used for the three AEGL tiers and the use of different data or studies for each tier, the data adequacy should be addressed separately for AEGL-1, -2, and -3 values. In addition to any discussion regarding the elements for evaluating key and supporting studies listed in this section of the TSD, the discussion should consider in general terms: (1) repeatability of experiments between laboratories, (2) consistency of data between experiments and laboratories, (3) types and number of species tested, and, (4) comparisons of the AEGLs with valid human and animal data.

A summary of the discussion in the TSD section "Data Adequacy and Research Needs" also should be included in the summary of the AEGL document and the derivation summary tables. The summary statements should address the adequacy of the data by AEGL tier.

2.4 DOSIMETRY CORRECTIONS FROM ANIMAL TO HUMAN EXPOSURES

When extrapolating from observed responses in animals to predicted human responses, the relationship between nominal exposure concentration and delivered dose to the target tissue is of concern. For inhaled toxicants, the target tissue is either a component of the respiratory system and/or other tissue or organ (systemic). A number of methods have been proposed to adjust for differences in the dose to target tissue in the respiratory system (EPA 1994) and those tissues located systemically (NRC 1993a; EPA 1994). The concern has been the lack of validated methodologies that would provide scientifically sound values for gases, vapors, and aerosols. That concern is particularly true when the methodology may predict levels for humans that may not be sufficiently protective. Both methodologies referenced above have not been validated for gases with experimental data, especially in the high-dose ranges required to produce toxicity with acute exposures. Another possible dosimetry correction, using the inhaled dose against the body weight raised to the 3/4 power, has support based on an analysis of chronic toxicity studies (EPA 1992). However, this adjustment may not be relevant for acute lethality studies (Rhomberg and Wolff 1998). Therefore, no dosimetry adjustments have been made to date by the NAC/AEGL Committee for attaining human-equivalent doses in the development of AEGLs for gases, vapors, and aerosols.

If AEGL values are developed for particulates, the methodology developed by EPA and validated with experimental data on particulate matter will be reviewed and applied on the basis of the individual material (EPA 1994). When specific data and validated models are available for chemicals inhaled as gases, a dosimetry correction will be made by the NAC/AEGL Committee.

2.4.1 Discussion of Potential Dosimetry Correction Methodologies for Gases

2.4.1.1 The Respiratory System As a Target Organ

The RfC (reference concentration) method for chronic exposure to gases was proposed by EPA (1994) as an approach to the dosimetry correction for effects on the respiratory system. This method has not been used by the NAC/AEGL Committee for the following reasons: (1) the RfC dosimetry corrections from animal to man are based on theoretical constructs that have not been confirmed and validated with experimental data; (2) some of the RfC assumptions are questionable and can have a significant impact upon the calculated dosimetry correction between animals and humans. Below is a discussion of two key examples and their impact upon the dosimetry adjustment. The assumptions are the requirement of uniform deposition in compartments and equivalent percent of deposition in animals and humans.

For Category 1 gases (highly water soluble and/or rapidly irreversibly reactive), the RfC method assumes that for each respiratory compartment (extrathoracic, tracheobronchial, and pulmonary), the deposition of chemical is equivalent throughout the compartment. This assumption fails to take into account major differences in anatomical structure and deposition (dose) as the gas, vapor, or aerosol progresses from proximal to distal regions within any one compartment. The dosimetric adjustment from rodent to man for the extrathoracic region predicts a 5-fold higher delivered dose to humans compared with rodents at equivalent exposures. However, a number of investigators have shown that treating the entire extrathoracic region as a single homogeneous compartment is incorrect. The use of sophisticated computational fluid dynamics computer modeling, correlated with analysis of patterns of lesions induced by chemical exposure, demonstrate that the degree of deposition of chemicals varies greatly in different extrathoracic regions in rats (Kimbell et al. 1993, 1997a,b) and the monkey (Kepler et al. 1998). Specific areas such as the olfactory epithelium will receive different regional doses in the rat and humans because of differences in surface area, susceptible location, and degree of ventilation (Frederick et al. 1998). A recent estimate of a dosimetric adjustment for vinyl acetate toxicity to the olfactory epithelium was performed using multiple compartments and a physiologically based pharmacokinetic model (PBPK). Bogdanffy et al. (1999) predicted that a time-adjusted exposure at 8.7 ppm in the rat would result in the same damage in a human exposed at 10 ppm. In this case, the application of the RfC method overestimates the risk to humans.

In the RfC method, the proportion deposited in each region for Category 1 gases is assumed to be the same in animals and humans. When the deposition is less than 100%, this assumption is incorrect when one considers a rodent breathing at 100 times a minute vs 15 breaths a minute for a human. The residence time for the chemical in a rodent lung is approximately 0.6 seconds and is approximately 4 seconds in a human, or about 6 times as long. All things being equal, the longer residence time in the human respiratory system will mean that the human extracts a greater percent of inspired chemical per breath than a rodent. Another factor to consider is that at high exposure levels a steady state can be rapidly achieved in which relatively little chemical is deposited in each breath so that the concentration becomes the determining factor.

Of concern is the fact that when dosimetry adjustments are made between rodents and humans for toxicity to the pulmonary region, the delivered dose to the human is predicted to be about 3 times less than the mouse for an equivalent nominal exposure concentration. Using this method in the absence of supporting empirical data could seriously underestimate human sensitivity. For example, at lethal concentrations, fluorine toxicity is due to pulmonary intoxication in all species tested (Keplinger and Suissa 1968). Further, the empirically derived LC_{50} values for the mouse, rat, rabbit, and guinea pig are essentially identical. However, the minute-volume-to-surface-area ratio for the pulmonary region of the guinea pig closely resembles the human. If the RfC dosimetry procedure were correct, the LC_{50} for the guinea pig should be 2-3 times higher than that observed for the rat and mouse, yet the empirical data were essentially identical for all three species. Using the RfC method to extrapolate a dosimetric correction to humans in this case would seriously underestimate the risk by a factor of 3 from the mouse data. This problem is compounded by the fact that the RfC method calls for the use of a lower interspecies uncertainty factor when the dosimetry correction is used.

2.4.1.2 Systemic Toxicity

Most systemic toxicants would fall under the definition of a Category 2 gas in the EPA methodology (EPA 1994). Category 2 gases are moderately water soluble and intermediate in their reactivity; thus, they would be distributed throughout the respiratory tract and absorbed readily into the bloodstream. In the case of Category 2 gases, the RfC dosimetry procedure predicts that the human receives a dose ranging from 6,000 to 50,000 times higher than a rodent (depending upon the species) for an equivalent exposure. These numbers do not appear to be biologically reasonable or scientifically credible. Because of the potential errors, the methodology for Category 2 gases has not been used. When a corrected methodology is published, it will be evaluated for use by the NAC/AEGL Committee.

For systemic toxicants, the NRC (1993a), proposed that dosimetry correction be conducted by adjusting for minute-volume-to-body-weight ratios. It is assumed for this calculation that 100% of the chemical or equal percentages of the chemical are absorbed. Given that assumption, the correction is a reasonable approach and may be valid for low concentrations of chemicals. Most animal-to-human extrapolation is done using mouse or rat data. Using certain typical minute-volume and body-weight parameters, it is possible to calculate an adjustment factor or multiplier to derive an equivalent dose in a human from animal data. The multiplier is approximately 6 for the mouse and 3.5 for the rat. Thus, if the exposure of interest in mice or rats is 100 ppm, an equivalent internal dose in humans would be predicted to be induced by exposure to 600 ppm and 350 ppm from these two species, respectively. Therefore, to induce an acutely toxic systemic effect in humans, people would have to be exposed to a concentration 6 times greater and 3.5 times greater than the nominal exposure required to induce the effect in mice or rats, respectively.

If less than 100% of the inspired chemical is absorbed with each breath, the human and animal would absorb a different fraction of the chemical in each minute (see discussion above). As the percent absorbed approaches 0, the multiplier would approach 1. In the example above, the multiplier for human dosimetry correction would decrease from 6 to 1 in the case of mice and 3.5 to 1 in the case of rats as the percent absorbed approaches 0.

AEGL-2 and AEGL-3 values represent relatively high exposure concentrations where absorption may not be complete. If the minute-volume-to-body-weight correction for dosimetry, which assumes 100% absorption, were used in these cases, the estimated human exposure equivalent to the rodent would be too high, leading to an underestimate of the toxicity and the derivation of AEGL values that are not protective of the human population.

Another approach to dosimetry correction may be that used by EPA when extrapolating from animal cancer bioassays to theoretical excess human cancer risk levels for lifetime exposures (EPA 1992). The cross species scaling factor used is based on an equivalence of $mg/kg^{3/4}/day$. There is reasonable scientific support for utilizing this approach based on an analysis of a number of multiple exposure studies across a number of animal species (EPA 1992). One might assume that the total amount of chemical inhaled is equivalent to the dose (NRC 1993a) and adjust that across species using the equivalence of mg/kg$^{3/4}$/day. However, Vocci and Farber (1988) point out that the power law of (body weight)$^{3/4}$ holds for the ventilation rate, such that on a weight-to-weight basis, the rat receives about 4 times the delivered dose of a human for the same exposure concentration. When this adjustment for breathing rate is combined with the adjustment for toxicity (EPA 1992), the two cancel each other out and one is left with the conclusion that equivalent exposure concentrations result in equivalent outcomes in animals and humans.

The situation is further complicated by an analysis of oral acute toxicity experiments by Rhomberg and Wolff (1998) using pair-wise comparisons of LD$_{50}$ values for different species for a large number of chemicals on the RTECS database. Their findings contrast with those of EPA (1992), which largely evaluated multiple exposure studies, in that the best correspondence of toxicity across species for LD$_{50}$ values was found when doses were expressed as milligrams per kilogram. This finding might argue for the NRC (1993a) recommendation to scale doses across species based on minute-volume-to-body-weight ratios. However, this conclusion would be based on an evaluation of oral toxicity studies, most of which were conducted by gavage. Bolus doses result in a high peak body dose, in contrast to the inhalation of a chemical over a number of hours with a more constant body burden over time. The question then becomes, does inhalation exposure on the order of hours mimic the toxic response seen with multiple exposures (EPA 1992) or the acute oral bolus doses used in the Rhomberg and Wolff (1998) analysis? If the former situation prevails, then the rationale by Vocci and Farber (1988) would argue for no dosimetry corrections being made. On the other hand, the latter case would argue for the use of the NRC (1993a) methodology.

2.4.2 Current Approach of the NAC/AEGL Committee to Dosimetry Corrections

Given the uncertainty surrounding this issue, and the fact that the use of no dosimetry corrections for gases across species would be the most conservative public-health approach, the NAC/AEGL Committee concluded it would not use dosimetry corrections across species. As the science surrounding this issue progresses, the NAC/AEGL Committee will re-evaluate that practice. If data are available on a chemical-by-chemical basis that scientifically support dosimetry corrections for gases in the development of AEGLs, they will be used.

As AEGLs are developed for particulates, the methodology developed by EPA and validated with experimental data on particulate matter will be reviewed and applied on the basis of the individual material (EPA 1994).

2.5 GUIDELINES AND CRITERIA FOR SELECTION OF UNCERTAINTY FACTORS TO ADDRESS THE VARIABILITY BETWEEN ANIMALS AND HUMANS AND WITHIN THE HUMAN POPULATION

2.5.1 Introduction

The variation in the toxicologic response of organisms to chemical exposures is well known. This variability is observed across species and among individuals within the same species. Lack of knowledge about the range of variability introduces uncertainties into any estimate of AEGL values based on biologic data. To account for known and unknown variability in response, the value derived from experimental data is adjusted by a value that reflects the degree of uncertainty. This value is referred to here and by most agencies and organizations as the uncertainty factor (UF). If an extrapolation is being made from animal data to humans, the total UF is a composite of an interspecies UF to account for possible differences between animal and human response to the chemical and an intraspecies UF to account for differences in response to the chemical within the human population. The intraspecies UF is needed to account for possible variabilities in response by "those at either extreme of age, those with poor nutritional status, those with preexisting diseases, such as certain heart diseases, that are fairly widespread in the general population, those with enhanced hereditary susceptibility, or those who are overexposed because of unusual physical exertion." (NRC 1993a, p. 88).

Interspecies and intraspecies UFs have been used in the development of "safe" or threshold exposure levels for chronic, noncancer toxicity by health organizations throughout the world. Examples include the acceptable daily intake (ADI) (Lu 1988; Truhaut 1991; Lu and Sielken 1991), the tolerable daily intake (TDI) or tolerable concentration (TC) (Meek et al. 1994; IPCS 1994), the minimal risk level (MRL) (Pohl and Abadin 1995), the reference dose (RfD) (Barnes and Dourson 1988; Dourson 1996), and the reference concentration (RfC) (EPA 1994; Jarabek et al. 1990). The importance of using distribution-based analyses to assess the degree of variability and uncertainty in risk assessments has been emphasized in recent trends in risk analysis. This will enable risk managers to make more informed decisions and better inform the public about possible risks and the distribution of those risks among the population (Hattis and Anderson 1999). These techniques can be used to assess variability from differences in individual exposure and susceptibility for specific risk assessments to reduce the uncertainty in estimating the real variability that exists in a population (Hattis and Burmaster 1994; Hattis and Barlow 1996).

The use of UFs in the development of AEGL values is designed to protect the general public, including susceptible subpopulations, from short-term exposures to acutely toxic chemicals. However, it is recognized that certain individuals, subject to unique or idiosyncratic responses, could experience adverse effects at concentrations below the corresponding AEGL value. "In the case of CEEL-2 [AEGL-2], UFs must be balanced against the inherent risk associated with actions, such as evacuation, that may be taken as a result of application of CEELs [AEGLs]. Large UFs, which may be appropriate with chronic exposure limits, such as PELs, may be associated with increased risk to the community in the application of CEEL-2 [AEGL-2]." (NRC 1993a).

For short-term exposure limits such as AEGLs, the NRC (1993a) recommended that UFs ranging from 1 to 10 generally be applied to account for interspecies and intraspecies variability. The selection of any UF should be based on the quality and quantity of data available for the chemical under consideration. Based on the data set at hand, and professional judgment, UFs between 1 and 10 are applied to account for interspecies and intraspecies variability in the derivation of AEGL values (NRC 1993a).

2.5.2 Background

The concept of "safety factors," more recently referred to as "uncertainty factors," is thought to have evolved from studies (Bliss 1935a,b) whose mathematical treatise of the statistical variations inherent in the calculation of the dosage-mortality curve indicated a range of uncertainty often found in toxicologic predictions (Calabrese 1983). Barnes and Denz (1954) observed that this same uncertainty may also occur for toxicologic predictions using data from animals. They suggested that food additives should show no adverse effect in test animals fed levels 100 times the levels used in food for humans. Although they considered this 100-fold margin of safety reasonable, they acknowledged at the time that there was no rigorous scientific basis for a value of 100.

Bigwood (1973) credits the Joint Food and Agriculture Organization-World Health Organization Expert Committee on Food Additives with developing the concept of "acceptable daily intake" (ADI) in the late 1950s. However, Lehman and Fitzhugh (1954) from the U.S. Food and Drug Administration (FDA) appear to be the first to suggest that ADIs be derived from chronic animal NOAELs expressed in milligrams per kilogram of diet through the use of a 100-fold factor. Bigwood (1973) criticized a WHO report published in 1967 on ADIs that states that an arbitrary factor of 100 has been widely accepted for determining maximum safe dietary levels but that the margin of safety may vary from 10- to 500-fold. Although acknowledging that the 100-fold safety factor (SF) is an approximation, Bigwood justified the value on the basis of (1) differences in body size of animals and humans, (2) differences in food requirements that vary with age, sex, physical exertion, and environmental conditions, (3) differences in water balance between the body and its environment, (4) differences in balance of hormonal functions (and its effect on food intake), and (5) differences in the susceptibility of various animal species to the toxic effect of a given additive. Further, he suggested that a SF slightly larger than 60 would cover the first four parameters and concluded that a SF of 100 would address all five parameters with an acceptable degree of validity. The WHO Expert Committee for pesticide residues used a similar approach (Lu 1979). Vettorazzi (1977, 1980) also supported the use of a 100-fold factor based on differences between animals and humans in susceptibility to toxicants, variations in response among humans, the size of the test population versus the exposed human population, the difficulty in estimating human intake, and possible synergistic actions of various chemicals in the human diet.

The FDA modified its original approach to ADI derivations using a 100-fold SF by accepting subchronic animal NOAELs with an additional 10-fold SF in the absence of chronic data (Kokoski 1976, as cited in Dourson and Stara 1983). The NRC (1977) recommended a similar approach with two significant changes: (1) the measurement of a NOAEL in milligrams per kilogram of body weight per day rather than milligrams per kilogram of diet, and (2) the use of a 10-fold factor when valid data on prolonged ingestion by humans were available. Also, several reports published during this time suggested that in certain instances the 100-fold SF could be divided into two 10-fold factors to describe separately the interspecies and intraspecies variabilities (Bigwood 1973; Klaassen and Doull 1980; Food Safety Council 1982).

On the basis of the NRC approach, EPA (1980) also recommended SFs for estimating ADIs for ambient water quality criteria. The ADIs were intended to protect individuals on the basis of chronic contaminant exposure. An additional SF of 1- to 10-fold was included to estimate NOAELs when only LOAEL data were available. Therefore, the FDA, NRC, and EPA all issued recommendations utilizing the 100-fold SFs with the additional agency-specific UFs described above.

While the original selection of SFs appears to have been rather arbitrary, Dourson and Stara (1983) introduced the concept that UFs (SFs) have empirical data to support their use for both intraspecies and interspecies adjustment. These workers analyzed the dose-response slopes of 490 rat LD_{50} studies reported by Weil (1972) and calculated the intraspecies adjustment factor required to reduce the dose level 3 standard deviations below the mean. This approach was used to predict the response of a susceptible subpopulation in a general population. They reported that an adjustment factor of 10-fold was adequate to reduce the response of a dose level killing 50% of the rat population to a level that would be lethal to only the most susceptible members (1.3 per 1,000) of the population for 92% of the chemicals studied.

Other studies evaluated for interspecies dose-response relationships by these workers, primarily between rats and humans, included metals (Evans et al. 1944, as cited in Dourson and Stara 1983), pesticides (Hayes 1967, as cited in Dourson and Stara 1983), arsenic and fluorine (Lehman and Fitzhugh 1954). Dourson and Stara (1983) concluded that although separate factors of 10-fold for interspecies and intraspecies response adjustments for chronic data appeared reasonable, more experimental data were needed. They believed that intermediate UFs of less than 10-fold for individual factors or less than 100-fold for combined factors could be used to estimate the ADI and may be developed on a logarithmic scale (e.g., 31.6 being halfway between 10 and 100 on a logarithmic scale).

Barnes and Dourson (1988) described EPA's approach and rationale to assessing noncarcinogenic health risks from chronic chemical exposure. EPA approach follows the general format as set forth by the NRC (NRC 1983). The conceptual difference between "safety" and "uncertainty" is discussed within the context of the terms SF versus UF and ADI versus RfD. The authors state that SF suggests the notion of absolute safety and that the ADI is generally and erroneously interpreted as a strict demarcation between what is "acceptable" and what is "safe" in terms of chronic exposure. Rather, the ADI represents an estimate of a xenobiotic exposure or daily dose where the probability of adverse effects is low but a level where the complete absence of all risk to all people under all conditions of exposure cannot be assured. Consequently, the RfD and UF terminology was developed and adopted by EPA. EPA considers the RfD to be an estimate (with uncertainty spanning perhaps an order of magnitude of a daily exposure to a human population, including susceptible subpopulations) that is likely to be without an appreciable risk of deleterious effects during a lifetime.

Dourson et al. (1992) conducted an analysis of chronic and subchronic toxicity data on 69 pesticides obtained from EPA's Integrated Risk Information System (IRIS) to determine the potential impact of missing studies on the quality of the RfD values derived. Certain of these data proved useful in determining interspecies variations in toxic response to long-term ingestion of a wide range of pesticides. The authors' analyses of 1- to 2-year studies indicated that the probability of the rat NOAEL for each of 67 pesticides exceeding the dog NOAEL by greater than 3.16-fold was 28%, and the probability of the rat NOAEL exceeding the dog NOAEL by greater than 10-fold was 10%. Also, the probability of the dog NOAEL in the same studies exceeding the rat NOAEL by greater than 3.16-fold was 19%, and the probability of the dog NOAEL exceeding the rat NOAEL by greater than 10-fold was 4%.

These data support the value of using UFs derived from data in developing RfDs and suggest that UFs between species may be significantly less than 10-fold for a wide range of structurally diverse chemicals.

Renwick (1993) considered the expression of toxicity to be the combined result of toxicokinetics (all processes contributing to the concentration and duration of exposure of the active chemical toxicant at the target tissue) and toxicodynamics (mode or mechanism of action of the active toxicant at the target tissue site). Therefore, he reasoned that since both toxicokinetics and toxicodynamics contribute quantitatively to the UF, it is necessary to subdivide each of the 10-fold UFs (interspecies and intraspecies) into these two components to effectively accommodate differences in contributions made by toxicokinetic and toxicodynamic factors. Hence, for any chemical, appropriate data may be used to derive a specific data-derived factor for that component. The overall interspecies and intraspecies UFs would subsequently be determined as the product of the known data-derived factor or factors and the "default" values for the remaining unknown factors. Renwick (1993) evaluated published data for parameters that measure interspecies differences in plasma kinetics (physiologic changes, differences in rates of absorption, biotransformation, and elimination) in laboratory animals and other parameters for interspecies differences in toxicodynamics, such as hematologic changes caused by exposure to organic and inorganic toxins and peroxisome proliferation in in vitro cell cultures. The author used a statistical analysis of variability in disposition rate of cyclamate and several pharmaceutical agents in healthy adults as examples for deriving the kinetic contribution to the intraspecies UF. The estimation of the intraspecies variability in dynamics was also based on drug responses in a similar human population. On the basis of these evaluations, Renwick proposed a division of the 10-fold UFs for both interspecies and intraspecies as 4-fold for differences in kinetics and 2.5-fold for differences in dynamics. Although Renwick conceded that the proposed subdivisions represent professional judgments, he stated that, based on consideration of appropriate data, the method was less arbitrary than using a 10-fold default to address each aspect.

Dourson et al. (1996) and Dourson (1996) summarized the status of UFs in noncancer risk assessments and discussed the use of "data-derived" UFs by various health organizations, EPA, and Health Canada. These authors presented data on interspecies and intraspecies variability of response that supported the use of lower UFs that are protective of human health. Dourson et al. (1996) also presented research and case studies from EPA and Health Canada risk assessments in which UFs other than a default value of 10-fold were used in the estimation of a RfD, RfC, TDI, or TC. The authors concluded that various organizations have begun to recognize that the default values currently used by risk assessors may in fact be inaccurate and overly conservative. As a result of the data available at that time, EPA and Health Canada began using UFs other than default UFs on a more regular basis. Dourson et al. (1996) provided the basis for concluding that the use of data-derived UFs for the development of RfDs should become the first choice and that only in situations in which there are truly inadequate data should the use of a 10-fold default factor be the first choice.

2.5.3 Considerations and Approaches to the Selection of UFs for Developing AEGLs

The concepts and the scientific data that support the use of UFs in human health risk assessment have progressed considerably during the past 50 years. The increase in knowledge of both interspecies and intraspecies susceptibility and variability, the toxicologic mechanisms of action, and the availability and evaluation of databases support the use of data-derived UFs.

The original 100-fold UFs (or SFs) were intended for use with food additives and were based on assumptions related to, or associated with, the oral route of exposure. Hence, considerations such as differences in food consumption, food requirements, water balance, and potential synergistic effects among various substances present in human food are not directly relevant to the consideration of inhalation exposure. Additionally, the original acceptable daily intake (ADI) was expressed in milligrams per kilogram of diet versus milligrams per kilogram of body weight, and application of the original 100-fold SF was generally accepted across chemicals with little or no evaluation of scientific data to support or reject the use of this value. Today, there is greater knowledge concerning bioavailability, and defined methods are available to evaluate susceptibility or variability in responses and selecting or deriving scientifically credible UFs.

Dourson and Stara (1983) introduced the concept that empirical data can support the use of UFs for both interspecies and intraspecies adjustment. This observation was followed by the publication of an analysis of the chronic and subchronic toxicity data obtained from EPA's Integrated Risk Information System (IRIS). Certain of those data proved useful in determining the extent of interspecies variations in response to long-term ingestion of a wide range of pesticides. More recently, the concept of data-derived UFs has been introduced (Renwick 1993; Dourson et al. 1996). Finally, the concept of dividing, evaluating, and quantifying separately the toxicokinetic and toxicodynamic factors from each of the interspecies and intraspecies UFs has been proposed (Renwick 1993).

One important consideration in the selection or derivation and use of UFs for the development of AEGLs is the nature of the toxicant and the exposure period. Much of the data, information, and emphasis to date on noncarcinogenic and nonmutagenic substances has addressed chronic effects from long-term or life-time daily exposures. Certain of the reports discussing the toxicokinetic and toxicodynamic factors as related to variability of response have drawn on carcinogenic or mutagenic mechanisms as a basis for scientific support. By contrast, the AEGL values address relatively high concentration, short-term exposures to threshold effects of acutely toxic chemicals. In attempting to draw on the scientific foundations upon which UFs are being selected for use in developing chronic guideline levels such as RfDs and RfCs, it is important to maintain an awareness of certain potential differences when considering acute guideline levels such as AEGLs. Responses to chronic exposures may be greater between species or between individuals as compared with responses to acute exposures. For example, the impact of individual differences in absorption, excretion, metabolism, rate of repair, or accumulation of unrepaired damage may be magnified through exposure to lower concentrations over extended time periods. The higher concentrations associated with acute exposure may tend to overwhelm existing defense mechanisms, possibly eliminating certain differences in response among species and among individuals within the same species. The higher concentrations associated with single exposures, together with the short-term nature of the exposure period, may reduce any differences in absorption, metabolism, and excretion of a substance, as well as differences in repair mechanism rates, and other factors. Hence, acute exposure to acutely toxic substances in some instances may reduce the variability in response between species and among individuals of the same species depending upon the mode of action of the material.

Based on the considerations presented above, the acceptance and use of default UFs based on chronic exposure data are carried out only after careful evaluation of chemical specific data for single exposures. However, the concepts, ideas, and approaches to developing UFs that have emanated from the chronic exposure studies of the past 10 years are valuable in the development of AEGLs and will be used as appropriate in the selection or derivation of UFs used in the AEGL program.

2.5.3.1 Interspecies UFs—Use in the Development of AEGL Values— Discussion

When data are insufficient to determine the relative susceptibility of animals in comparison to humans, a UF of 10 has been used by EPA, ATSDR, Health Canada, WHO, the International Programme on Chemical Safety (IPCS), and Rijksinstituut voor Volksgesondheid en Milieu (RIVM) when developing the equivalent reference doses for chronic exposure to chemicals (Dourson et al 1996). When extrapolations are made from animals to humans based on milligrams per kilogram of body weight, the factor of 10-fold is usually adequate to account for differences in response. Dourson and Stara (1983) found that a factor of 10 accounted for many of the animal-to-human differences observed when the dose was adjusted for differences between human and animal body weights and body-surface areas.

Brown and Fabro (1983) compared the lowest effective dose to cause teratogenicity in animals (mouse, rat, rabbit, cat, and monkey) and humans for eight chemicals (methylmercury, diethylstilbesterol, methotrexate, aminopterin, polychlorinated biphenyls (PCBs) thalidomide, phenytoin, and alcohol). The LOAEL ratios ranged from 1.8 to 50 with a geometric mean of 7. Humans were generally more susceptible on an administered oral-dose/body-weight basis but by less than an order of magnitude. This analysis is complicated by the fact that the criteria and confidence in determining the lowest effective dose are not discussed, and the chemicals may represent potent developmental toxicants in humans since their confirmed teratogenicity in humans represented the basis for their selection. The potency estimates in humans may represent only the sensitive part of the distribution of human response to exposure. The animal response dose may be closer to the mean response level, and therefore presents a higher LOAEL for the species used in this comparision. However, the retrospective nature allows the choice of the most susceptible animal species. In most instances the animal database is incomplete. Thus, this analysis may represent an incomplete spectrum of results in which humans appear more susceptible than animals to certain developmental toxicants.

Renwick (1993) subdivided the interspecies and intraspecies UFs into two components to address toxicokinetics and toxicodynamics separately. Although supporting data for this concept stem from chronic animal feeding studies and in vitro cell cultures, the concept of considering the kinetics and dynamics separately across species has relevance to UFs for AEGLs. Renwick proposed specific quantitative values of 4-fold and 2.5-fold for the kinetic and dynamic components, respectively. Although this approach has merit, the NAC/AEGL Committee does not make such a precise quantitative differentiation. To date the NAC/AEGL Committee uses only general information on the kinetic and dynamic components of toxicity to adjust the interspecies UF (e.g., from 10 to 3 or 1). This approach is also consistent with the recommendation by Dourson et al. (1996) to use data-derived UFs when appropriate data are available. This approach is in keeping with EPA's general approach in the development of RfDs. For example, in the case of Aroclor 1016, the default interspecies UF of 10 was reduced to 3 because of the similarity with which monkeys and humans respond to PCBs (toxicodynamics), and metabolize PCBs (toxicokinetics), and the physiologic similarity (toxicokinetics) between the two species (EPA 1996b).

Comparisons of the current approach to determine UFs for AEGLs with other short-term exposure limits have not altered the practice of the NAC/AEGL Committee. In the development of emergency exposure guideline levels (EEGLs) by the NRC (NRC 1986), a factor of 10-fold was used for interspecies extrapolation.

The NRC *Guidelines for Developing Spacecraft Maximum Allowable Concentrations for Space Station Contaminants* (SMACs) states that UFs between 1- and 10-fold are used for each source of uncertainty (NRC 1992a). The sources include intraspecies (human) response variabilities, interspecies variabilities, the extrapolation of a LOAEL to a NOAEL, and the extrapolation from an inadequate or incomplete data base. For 1-h SMACs, the NRC used an overall (combined intraspecies and interspecies) UF of 10-fold when only animal data were available or when the route of human exposure differed from the study. However, the population for which SMACs is intended (astronauts) does not include infants, children, the elderly, or the infirm and is, therefore, a more homogeneous and healthier subpopulation.

The NRC (1993a) recommended the use of an interspecies UF within the range of 1- to 10-fold to account for differences between animals and humans. The guidance suggests that the UF should be based on the quality of the data available. In this regard, the NAC/AEGL Committee evaluates data on a chemical-by-chemical basis, considers the weight of evidence, and uses scientific judgment in the selection of interspecies UFs. As data are available, the NAC/AEGL Committee uses data-derived interspecies UFs.

Information bearing on the toxicokinetics and toxicodynamics of the chemical under consideration, as well as structurally related analogues or chemicals that act by a similar mechanism of action, will be used to derive an appropriate interspecies uncertainty factor that may range from 10 to 3 or 1. In the absence of information on a subject or analogous chemical to set data-derived UFs, the use of a default UF of 10 is considered to be protective in most cases. As always, all information on the chemical, its mechanism of action, structurally related chemical analogues, and informed professional judgment will be used when determining appropriate UFs and evaluating the resultant AEGL values.

2.5.3.2 Interspecies UFs—NAC/AEGL Committee Guidelines

The NRC (1993a) provides guidance on approaches to selecting the most appropriate species for deriving AEGL values. General guidelines followed by the NAC/AEGL Committee to select UFs are presented below. In each section, there is a list of questions that should be addressed to support the rationale for the UF used. The guidelines are organized into categories for convenience. However, more than one guideline may be applied to the selection of any one UF.

2.5.3.2.1 Small Interspecies Variability or Most Appropriate Species Used

In cases in which the interspecies variability is small (e.g., within a factor of 3), the most susceptible species is selected, or a species whose biologic response to the substance in question is closely related to humans is selected, the interspecies UF is typically 3. It should be noted that in those cases in which the mode of action can be identified and there is evidence that it is not expected to vary significantly among species, the UF is generally 3.

The rationale for the selection of a UF should include the following:

1. The species tested.
2. The toxicologic endpoint used for the AEGL derivation.
3. The qualitative and quantitative range of responses of the species tested.
4. Discussion of why the species and study chosen was the most appropriate.
5. Discussion of the variability among studies with the same species or among strains.

2.5.3.2.2 Most Susceptible Species Not Used

In instances in which the most susceptible species is not used, a UF of 10 is generally used.

The rationale for the selection of a UF should include the following:

1. The species tested.
2. The toxicologic endpoint used for the AEGL derivation.
3. The qualitative and quantitative range of responses of the species tested.
4. Discussion of why the most susceptible species was not used and/or why the less-susceptible species was selected.

2.5.3.2.3 Mechanism or Mode of Action Is Unlikely to Differ Among Species

If evidence is available indicating that the mechanism or mode of action, such as direct-acting irritation or alkylation, is not expected to differ significantly among species, an interspecies UF of 3 is generally used.

The rationale for the selection of a UF should include the following:

1. A description of the mechanism of action.
2. A discussion of why the mechanism of action is unlikely or likely to differ.
3. Is bioavailability, metabolism, detoxification, elimination likely to be an issue?

2.5.3.2.4 Mechanism or Mode of Action Is Unknown

In cases in which the mechanism or mode of action is unknown, insufficient data on differences between species are available, or there are likely to be substantial (but inadequately quantified) differences in metabolic and physiologic response between species, an interspecies UF of 10 is applied.

The rationale for the selection of a UF should include the following:

1. Description of the toxicologic effects observed.
2. Description of the range of uncertainty in toxicologic response and how that relates to this assessment.
3. Discussion of what is known and unknown about the mechanism or mode of action.
4. Discussion of the extent of data available among species.

2.5.3.2.5 Large Variability in Response Between Species

When there is a high degree of variability among species or strains of animals or experiments that cannot be explained adequately, an interspecies UF of 10 is applied.

The rationale for the selection of a UF should include the following:

1. Description of the response.
2. Discussion of the differences or similarities in pharmacokinetic parameters (absorption, metabolism, detoxification, elimination) among species.
3. Discussion of the range of dose-dependent responses of the species tested and the qualitative and quantitative aspects of the data.

2.5.3.2.6 Humans More Susceptible Than Animals

When published data show humans are more susceptible than animals, an interspecies UF of 10 is used unless published results demonstrate otherwise.

The rationale for the selection of a UF should include the following:

1. Description of the toxicologic endpoints for which humans and animals show differential susceptibility.
2. Discussion of the factors where humans are thought to be more or less susceptible than animals.
3. State what species were tested.
4. Discussion of the range of response of the species tested. This discussion should address qualitative and quantitative aspects of the data.
5. Discussion of human factors that account for unique susceptibility compared with test animals.

2.5.3.2.7 Inadequate Data

The UF for interspecies response adjustment is 10 when there are inadequate data or insufficient information about the chemical or its mechanism of action to justify an alternative UF.

The rationale for the selection of a UF should include the following:

1. Discussion of the inadequacy of the data that are the basis for a UF of 10. For example, the analysis may depend upon data collected in only one species, high variability of response, or uncertainties in exposure measurement. The statement may point to data gaps that could be filled where the need exists.

2.5.3.2.8 A Selected UF Applied to Animal Data Driving the AEGL-2 or -3 to a Value Tolerated by Humans Without Lethal or Serious Adverse Effects

When the application of an interspecies UF of 10 reduces the AEGL-3 value (the threshold for lethality) or the AEGL-2 value (the threshold for irreversible or disabling effects) to an exposure concentration that humans are known to tolerate without adverse effect, the interspecies UF is reduced to 3 or 1.

The rationale for the selection of a UF should include the following:

1. Citations and explanations of the human data and how it relates to the AEGL value derived with a UF selected on the basis of the existing guidelines.

2.5.3.2.9 A Multiple Exposure Study Used to Set the Level

In cases in which an AEGL value is derived from a multiple exposure study, because the data set for a single exposure is lacking, the multiple exposure data are considered an inherently conservative estimate. This conclusion is based on the observation that a biologic organism often demonstrates greater tolerance to a single exposure compared with multiple exposures at the same or similar levels to the same chemical. If the adverse effect identified in the multiple exposure study is cumulative for the AEGL of concern, the interspecies UF used to adjust the multiple exposure animal data may be reduced. Careful judgment should be used when making this assessment. If a chemical is cleared very rapidly, or if there is evidence of short-term adaptation, or if there is evidence that the concentration causing the effect does not vary with duration or number of exposures, then the animal may be able to sustain repeated insult at a level close to a single acutely toxic exposure. In those instances, the reduction of the UF based on multiple exposures versus a single exposure may not be justified.

The rationale for the selection of a UF should include the following:

1. A description of the study.
2. Discussion of the empirical clearance rate and other toxicokinetic properties of the chemical. For example, does the concentration causing the effect vary significantly with duration or number of exposures?

2.5.3.2.10 Selection of the NOAEL

The highest NOAEL is generally selected in instances in which acceptable direct or supporting data exist. In such cases, a UF of 3 or 10 is used, depending on the available data. Although UFs are typically applied, in certain cases, multiple species have similar NOAELs, the NOAEL selected is substantially below the NOAELs reported for other species, or there is a high degree of confidence that the animal model tested is a sensitive surrogate for humans or is more sensitive than humans. In such cases, a UF of 1 might be used. In all cases, the NOAEL represents the highest exposure level from animal data in which the effects used to define a given AEGL tier are not observed.

2.5.3.3 Intraspecies UFs Used in the Development of AEGL Values— Discussion

Intraspecies UFs are used to address the variability in biologic response that exists within a human population exposed to a toxic agent. Their use represents an important step in the AEGL development methodology and is designed to account for the differences that can occur within the general population.

The NRC guidelines for developing community emergency exposure levels (CEELs) state that the exposure levels are "designed to protect almost all people in the general population...." (NRC 1993a). The NRC guidelines state that although the levels "are designed to protect 'susceptible' individuals, some hyper-susceptible individuals might not be protected...." That distinction is based on the premise that CEELs must be set low enough to protect the general population but must also be set at levels that minimize the public health and safety risks associated with response to chemicals as a result of rare or exceptional circumstances. Consequently, the AEGL values may not be expected to necessarily protect certain individuals with unique or idiosyncratic susceptibilities. This consideration is clearly communicated in the NAC/AEGL Committee's definition of the AEGLs.

When data are insufficient to determine the relative susceptibility of individuals in a human population exposed to a specific chemical, a default UF of 10 has been used by EPA, ATSDR, Health Canada, IPCS, and RIVM when developing the equivalent reference doses for chronic exposure to chemicals (Dourson et al. 1996). This value of 10 is generally applied to the NOAEL. A number of studies have tried to address the issue of the reasonableness or validity of this UF. Under ideal circumstances, an analysis would provide information on the ratios of the experimentally observed NOAELs for different human groups within a population for a wide range of defined exposures to chemicals. Groups would be identified on the basis of biochemical or physiologic differences that might cause members of the group to respond to chemical exposure in a fundamentally different manner—either quantitatively or qualitatively. Sample sizes would be large and include a wide variety of genetic backgrounds. Such examples would include differences among newborns, infants, children, adults, the elderly, the infirm, and those compromised by illness, including those with asthma. The NOAELs also would represent a distinct relationship between dose level and response. These data would encompass all variables due to the toxicokinetics and toxicodynamics factors. Such data are not available, even in carefully controlled, double-blind clinical trials for new therapeutic drugs. However, surrogates have been developed that provide information on the reasonableness of the choice of the intraspecies UF of 10 or less. This approach is referred to as the use of data-derived UFs.

Dourson and Stara (1983) analyzed the slopes of LD_{50} gavage studies in 490 adult rats reported by Weil (1972). They calculated the intraspecies adjustment factor required to reduce the dose 3 standard deviations below the median LD_{50} response using a probit, log-dose analysis. This gives a z value of 0.4987 from the mean or a calculated response of 1.3/1,000 (Spiegel 1996). This value was used to predict the response of a susceptible subgroup in the population. An adjustment factor of 10 was adequate to reduce the response from a dose killing 50% of the animal population to a level that would kill only the most susceptible members of the inbred rat population in 92% of the chemicals studied. These data support the contention that a 10-fold UF is adequate in many instances to account for intraspecies differences in response to acute exposures. However, in some instances this UF may not protect the more susceptible members of the population. The extrapolation reported here represents a measure of 3 standard deviations from the median response data points. Statistically, an extrapolation of 3 standard deviations from the mean includes more than 99% of the population in question, or approximately 999 individuals out of a population of 1,000. The extrapolation of 3 standard deviations as performed by Dourson and Stara (1983) includes a similar proportion of the population in question, 998.7 of 1,000. It is interesting to note that the Fowles et al. (1999) analyses of inhalation toxicity experiments revealed that for many chemicals the ratio between the LC_{50} and the experimentally observed nonlethal level was on average a factor of approximately 2, the 90th percentile was 2.9, and the 95th percentile was 3.5. There was a range of ratios from 1.1 to 6.5. Therefore, the use of a UF of 3 with a NOAEL for lethality can achieve the same reduction in acute lethality as that reported by Dourson and Stara (1983). The 490 LD_{50} studies with rats were undoubtedly based on a wide range of chemical substances exhibiting many different toxicologic mechanisms. Hence, the variability due to chemical-specific properties was included in this evaluation and was accounted for by an adjustment factor of 10 in 92% of the chemicals tested. This type of statistical analysis makes the untested hypothesis that the slope of the dose response was the same in the experimental dose range and at the untested tails of the experiment. It also reflects the response in a homogeneous (inbred) adult animal population and does not measure the difference in values between potentially susceptible subgroups, such as adult vs newborn.

A number of authors have presented data and analyzed adult/newborn LD_{50} ratios to assess the differential susceptibility of young and adult animals. Done (1964, as cited in NRC 1993b) compiled LD_{50} ratios between immature and mature animals. For 34 of 58 chemicals, Done found that the immature animals were more susceptible than the adults, and for 24 of 58 chemicals, the adults were more susceptible than the immature animals (NRC 1993b). A similar compilation of newborn/neonate and adult LD_{50} ratios for rat and mouse was done by Goldenthal (1971) on data submitted to FDA in drug applications. This compilation included a broad range of chemicals, such as analgesics, bronchodilators, central-nervous-system depressants and stimulants, antidepressants, and tranquilizers. NRC (1993b) analyzed these data and found that about 225 of the compounds were more toxic to neonates and 45were more toxic to adults. Almost all of the age-related differences from the Done (1964, as cited in NRC 1993b) and Goldenthal (1971) data collections were within a factor of 10 of each other, and most of the ratios were within a factor of 3 (NRC 1993b). Sheenan and Gaylor (1990) analyzed adult/newborn LD_{50} ratios for 238 chemicals. The median ratio of the LD_{50} values between age groups was 2.6. Approximately 86% of the ratios were less than 10, indicating that this factor is adequate to account for differences in response to chemical exposure between adult and young in most cases but may be insufficient for 14% of the cases. In these studies, the comparison was made from the median response.

Another indirect approach to quantify biologic uncertainty is to measure the observed variability in human populations. Calabrese (1985) examined a number of parameters related to toxicokinetics (metabolism, binding of chemicals to protein and DNA, and activity levels of enzymes). In studies that included between 10 and 349 subjects, Calabrese concluded that generally 75-95% of the population fell within a range of 10-fold. However, the author's conclusion was based on the supposition that the 10-fold factor was to account for the total range of human variability as opposed to the range from an experimental no-observed-effect level to the most susceptible person. In a similar study, Hattis et al. (1987) evaluated toxicokinetic parameters in 101 data sets (five or more healthy adults) on 49 chemicals (primarily drugs). They found that 96% of the variation was within a factor of 10. However, this analysis also measured the total range of human variability. These analyses measured the range of responses for toxicokinetic parameters and give some indication of the variability in an adult population only and not in a potentially susceptible subpopulation. They do not measure how far the tail for response goes beyond the lowest dose in the population measured, nor the response of different populations. Another consideration is the fact that these data represent measures of toxicokinetic variables that may not directly reflect the threshold of toxicologic response to chemical exposure.

Ideally, one would like to be able to compare NOAELs observed in an experiment with the tail of the NOAEL distribution to assess the actual frequency of response in the total human population when the intraspecies UF of 10 is applied and obtain a measure of the susceptible person. Determining the experimental NOAEL is fraught with problems of sample size and dose selection. The response of the susceptible subpopulation at a dose 10-fold lower than the experimental NOAEL will never be known. Hattis et al. (1999) performed statistical modeling analyses designed to determine the efficacy of applying the intraspecies UF of 10 to a NOAEL. The authors statistically analyzed clinical studies on humans that measured parameters related to toxicokinetics and toxicodynamics. The studies had at least five subjects each and included approximately 2,700 data points for the toxicokinetic endpoints. They demonstrated that the population distribution of the data were lognormal in the data region and assumed that they were lognormally distributed out to the extreme tails. From the data, and assuming a lognormal distribution, they calculated the dose required to produce an incidence in 5% of the population. This is essentially an experimental NOAEL that is divided by the intraspecies UF when a risk assessment is performed. The dose at the 5% incidence level was divided by 10 and the response at that dose calculated, assuming a lognormal distribution of data, to the extreme tails. This approach was used to assess the response rate when a 10-fold UF is applied to a NOAEL. They found that "acting by itself, a 10-fold reduction in dose from a 5% effect level could be associated with effect incidences ranging from slightly less than one in ten thousand for a median chemical/response to a few per thousand for chemicals and responses that have more human interindividual variability than 19 out of 20 typical chemicals/responses." The analysis did not include susceptible subpopulations, so the variability seen could be greater. This type of analysis assumes a lognormal distribution of the data to the extreme tails. It does not allow for a threshold that is generally assumed to be true for noncancer effects. Thus, the calculated response at doses 10-fold less than the 5% response level may be overly conservative. There are no data, human or animal, that far out in the tail of the distribution curve. The analysis by Hattis et al. (1999) indicates that a human intraspecies UF of 10 would be protective of susceptible subpopulations and may be overly conservative in many instances.

Another approach to measuring variability among different groups of a human population is to compare maximum tolerated doses (MTDs) or effect levels between groups. Reports comparing the MTDs of chemotherapeutic agents in child and adult cancer patients indicate that most of the substances studied were tolerated as well and, in many instances, tolerated better by children than by adults when the dose was expressed as milligrams per kilogram of body weight or milligrams per square meter (Glaubiger et. al. 1982; Marsoni, et. al. 1985). In those instances in which children demonstrate a greater response at equivalent dose to these substances, the differences were less than a factor of 2-fold. Although MTDs are not entirely a precise measure of a toxicologic threshold, they represent a credible parameter by which relative toxicities between groups can be measured in humans. It is important to acknowledge that although the substances studied represent a diverse group of chemical classes, these substances exhibit similar mechanisms of cytotoxicity. Therefore, the results observed cannot be applied to a large number of other chemicals with different mechanisms of action. In addition, only MTDs were reported, not the variability within each group in response to the drugs. Thus, this type of study gives a measure of response among groups within a population but not the variability within each group.

Other studies regarding differences in susceptibilities among specific groups in humans to various anaesthetic gases have been reported. These studies indicate children, particularly infants, are more resistant than adults to the effects of various volatile anesthetics (Gregory, et. al. 1969; Stevens, et al. 1975; Lerman et. al. 1983; LeDez and Lerman 1987; Katoh and Ikeda 1992; Chan et al. 1996). The susceptibility of individuals of different ages has been extensively studied in the anesthesia literature in which the concentrations of various anesthetic gases in the lung, producing "anesthesia" (i.e., lack of movement), have been measured. The results are usually reported as the minimum alveolar concentration (MAC) that produces lack of movement in 50% of persons exposed to that concentration. Occasionally, the ED_{95} (the alveolar concentration that prevents movement in 95% of those exposed) is also reported. MACs for several anesthetic gases have been measured as a function of age. The results consistently show a pattern with maximal susceptibility (lowest MAC values) in newborns, particularly prematures, pregnant women, and the elderly. The least susceptibility (highest MAC values) occurs in older infants, toddlers, and children as compared with adults. The total range of susceptibility was 2- to 3-fold. Many organic solvents for which AEGLs are developed can also produce anaesthesia in humans at high doses. As previously stated, this type of study gives a measure of response between groups within a population but not the variability within each group.

Intraspecies UFs are used to address the variability in biologic response that exists within a human population exposed to a toxic agent. Their use is designed to account for the range of responses to exposure by individuals within the general population. As the studies above demonstrate, a UF of 10 is adequate to account for variability in the majority of cases and a factor of 2-3 is often adequate.

It has been proposed that data on the differences in kinetics and dynamics be used to modify the UFs from defaults of 10 (Renwick 1993; Dourson et al. 1996). Renwick (1993) proposed dividing interspecies and intraspecies UFs into two components. Toxicity is considered to be the combined function of toxicokinetics (all processes contributing to the concentration and duration of exposure of the active chemical toxicant at the target tissue) and toxicodynamics (mode or mechanism of action of the active toxicant at the target tissue site). If data are available on the differences between or within species on one or both of these two processes, then it should be possible to reduce the total UF by developing a data-derived UF. This approach has in fact been taken by EPA in the examples below. EPA (1996b) reduced the UF of 10 to 3, because data from specific susceptible subpopulations were unavailable.

In the case of methylmercury, toxicodynamic data were used to reduce the intraspecies UF to 3 (EPA 1995b). The RfD was based on a benchmark dose computed as the lower 95% confidence limit on the 10% increase over the background for human childhood neurologic abnormalities (this level has been used to represent the NOAEL) in the susceptible subpopulation (the developing fetus). Therefore, the default intraspecies UF of 10 was reduced to 3. Since the susceptible subpopulation had been identified, the toxicodynamic part of the UF had been addressed. However, variability due to toxicokinetics was maintained with the use of the 3-fold UF.

For styrene, the default intraspecies UF of 10 was reduced to 3 in the calculation of the RfC value, because the lower 95% limit of the exposure extrapolation for a NOAEL in a human cross-sectional study was used and the biologic exposure index had been shown to account for variation in pharmacokinetic and physiologic measures, such as the alveolar ventilation rate (EPA 1993).

In the absence of information to justify data-derived UFs, a UF of 10 is considered to account for intraspecies variability in most cases. When information is available about the response of a susceptible subpopulation (e.g., mechanism of action in different species and/or subgroups within an exposed population, toxicokinetic data, or toxicodynamic data), those data are factored into the development of a data-derived UF, which may vary between 10 and 1. All information on the chemical, its mechanism of action, structurally related chemical analogues, a discussion of the weight of evidence, and informed professional judgment are used when determining UFs.

2.5.3.3.1 Range of Susceptibility

The definition and intended application of AEGL values make distinctions between susceptible and "hypersusceptible" individuals. It is important to characterize these two terms and the potential subpopulations they may represent for purposes of UF selection. It is also important to distinguish between these two populations for purposes of risk communication to emergency planners, emergency responders, and the public.

Individual susceptibility within a population will vary according to individual determinants and the specific properties of a given chemical. The origins of susceptibility are multifactorial and distributed across populations. According to the U.S. Presidential/Congressional Commission of Risk Assessment and Risk Management, "Genetic, nutritional, metabolic, and other differences make some segments of a population more susceptible than others ... susceptibility is influenced by many factors" (PCCRARM 1997). The factors are based on intrinsic and acquired differences among individuals and may include age, gender, genetic factors, ethnicity and race, and quality of life and life-style considerations. The latter considerations may be further classified as pre-existing illnesses, prior exposures, nutritional status, personal behavior (e.g., occupation, smoking, alcohol use, and obesity), and socioeconomic factors. The NRC also characterizes such determinants: "[S]ome of the individual determinants of susceptibility are distributed bimodally ... other determinants seem to be distributed more or less continuously and unimodally" (NRC 1994).

Hypersusceptibility describes extreme examples of responses. It may represent biologic reactions that are unique, idiosyncratic, and stem from determinants that are generally discontinuous with, and lay outside of, the range of distributions expected for the general population.

The determination of susceptibility entails the presence of observable changes in biochemical or physiologic processes reflecting dose-response relationships unique to a chemical (e.g., sulfur dioxide) or class of chemicals (e.g., acid aerosols). Susceptibility and hypersusceptibility are not meaningful concepts outside of the context of specific exposures. "Dose-response relationships are chemical-specific and depend on modes of action; people are not hypersusceptible to all kinds of exposures" (PCCRARM 1997).

Susceptibility and hypersusceptibility may reflect transient rather than permanent states. For example, infants are susceptible to some chemicals (e.g., ingested nitrates and nitrites as a result of relatively high gastric pH), but lose that susceptibility as they mature. Susceptible populations may also experience transient periods of hypersusceptibility. For example, asthmatics represent 5-10% of the general population and can be more susceptible than nonasthmatics to challenge by respiratory irritants. Moreover, at any given time some asthmatics may be suffering acute asthmatic attacks, which might lead to a hypersusceptible condition, just prior to an irritant exposure. Because of the transient condition, these individuals might not be accounted for in the published AEGL values. Similarly, otherwise normal individuals may suffer transient periods of hypersusceptibility during periods of illness. For example, following very severe, acute respiratory infections, many nonasthmatic individuals will experience several weeks or more of bronchiolar hyperreactivity and bronchospasm following nonspecific exposure to respiratory irritants. This condition can be considered an example of transient hypersusceptibility. In general, since there is little or no information regarding the responses of transiently hypersusceptible individuals to chemical exposures, the corresponding AEGL values might not be protective for this group.

During the past 15 years, a wide range of symptoms and complaints in patients thought to be related to extreme sensitivity to low levels of diverse and often nonquantifiable chemical exposures have been reported. This syndrome has been referred to as "multiple chemical sensitivity" or MCS (Cullen 1987). MCS has been characterized as the heightened, extraordinary, or unusual response of individuals to known or unknown exposures whose symptoms do not completely resolve post-exposure or whose sensitivities seem to spread to other chemicals (Ashford 1999). The syndrome is thought by Ashford to be a two-step process with an initial acute exposure to high concentrations of a substance and the subsequent triggering of symptoms at extraordinarily low levels of exposure to the same substance or different substances. He believes that repeated or continuous lower level exposures may also lead to the same type of sensitization. Ashford and Miller (1998) also postulate that this sensitivity may be the consequence of a variety of disease processes resulting from "toxicant-induced loss of tolerance"—described as "a new theory of disease providing a phenomenologic description of those disease processes."

In response to the increasing public demand for governmental attention to a problem frequently identified as MCS, the Environmental Health Policy Committee (EHPC) of the U.S. Public Health Service (PHS) formed the Interagency Workgroup on MCS in 1995 to address this issue (Mitchell 1995). The workgroup's charge was to review the scientific literature on MCS, consider the recommendations from various expert panels on MCS, review current and past federal actions, and make recommendations to policy-makers and researchers at government agencies concerned with evaluating public-health issues that might relate to MCS-like syndromes. The workgroup comprised scientists from the U.S. federal agencies, including, ATSDR, DOD, DOE, Department of Veterans Affairs, National Center for Environmental Health, Centers for Disease Control and Prevention, National Institute of Environmental Health Sciences, National Institutes of Health, and EPA. The original draft report was peer reviewed by 12 independent experts in occupational and/or environmental medicine, toxicology, immunology, psychology, psychiatry, and physiology. A Predecisional Draft Report was issued for public comment on August 24, 1998 (Interagency Workgroup on Multiple Chemical Sensitivity 1998). Although a final report has not yet been issued, the draft report concluded that MCS remains a poorly defined problem. The experts disagree on possible causes (e.g., physical or mental), and the sufferers complain of a wide range of symptoms (not associated with any "end-organ" damage) that may result from a disruption of homeostasis by environmental stressors.

In addition to the EHPC Interagency Workgroup on MCS, the NRC (NRC 1992b,c), professional organizations (ACOEM (see McLellan et al. 1999); AAAI 1986; AAAAI 1999), and others (Kreutzer et al. 1999; Kipen and Fiedler 1999; Graveling et al. 1999) have attempted to address this issue. Despite these attempts, the diagnosis, treatment, and etiologic assessment of MCS has remained a troublesome medical and social concern for individuals, physicians, government, and private organizations (McLellan et al. 1999). No consensus has yet been reached for a case definition (Mitchell 1995; McLellan et al. 1999); Graveling et al. 1999), diagnostic methods (Mitchell 1995; AAAAI 1999; McLellan et al. 1999), or treatment (AAAAI 1999). Further, despite extensive literature on the existence of MCS, "there is no unequivocal epidemiologic evidence; quantitative exposure data are lacking; and qualitative exposure data are patchy" (Graveling et al. 1999). Although most researchers contend that symptoms characteristic of chemical sensitivities exist, they agree that symptoms may be exaggerated and may be "differentially precipitated by psychosocial events or stress, or by different physical or chemical exposures" (Ashford 1999). All researchers and clinicians familiar with the problem agree more work must be done to understand the unexplained symptoms that are attributed to MCS (Kipen and Fiedler 1999).

The American College of Occupational and Environmental Medicine (ACOEM), the American Academy of Allergy, Asthmatics, and Immunology (AAAAI) and the International Programme on Chemical Safety (IPCS) have all recommended that the term "idiopathic environmental intolerance" be used to replace the term MCS (McLellan et al. 1999; IPCS 1996; AAAAI 1999). These authors believe that the term MCS incorrectly implies that the condition affects the immune system and that chemical exposure is its sine qua non (McLellan et al. 1999). No immunologic dysfunction has been identified in these patients (Graveling et al. 1999; AAAAI 1999). Further, they concur with other prominent medical organizations in maintaining that evidence does not exist to define MCS as a distinct clinical entity (McLellan et al. 1999).

While some clinicians hold that MCS occurs as a result of environmental exposures, mechanism(s) by which that may take place have not been proven scientifically. No single widely accepted test of physiologic function can be shown to correlate with observed symptoms (Interagency Workgroup on Multiple Chemical Sensitivity 1998; Brown-DeGagne and McGlone 1999; AAAAI 1999; McLellan et al. 1999). Immunologic, allergic, neuropsychologic, and traditional psychiatric disorders have all been postulated to cause MCS, but to date, they have not been supported by well-designed studies (Interagency Workgroup on Multiple Chemical Sensitivity 1998; McLellan et al. 1999; Brown-DeGagne and McGlone 1999).

As a result of the considerations presented here, it is not believed that MCS represents a viable scientific basis for developing AEGL values, including further adjustments for susceptible subpopulations, at the present time. However, the NAC/AEGL Committee recognizes the need for scientific research on this proposed syndrome that may help explain and describe its features, enable scientifically valid approaches to hazard or risk assessment, and define appropriate clinical interventions. Also, the committee considers all new data or information that is scientifically credible and relevant to the development of AEGL values.

2.5.3.3.2 Selection of Intraspecies UFs

To meet the AEGL definitions that protect susceptible subpopulations but not necessarily hypersusceptible subpopulations, the NAC/AEGL Committee evaluates two separate considerations regarding susceptibility. First, evidence is reviewed to attempt to distinguish "susceptible" from "hypersusceptible" subpopulations for each chemical of concern. Second, estimation of the range of response variability is carried out in the general population that includes susceptible (but not necessarily hypersusceptible) subpopulations and selection of appropriate intraspecies UFs for development of the AEGL values.

2.5.3.3.3 Distinguishing Susceptible and Hypersusceptible Subpopulations

A clear distinction between susceptible and hypersusceptible subpopulations in all cases for all chemicals is not achievable with the clinical and toxicologic information available to date. However, the NAC/AEGL Committee has identified specific categories and subpopulations that may be considered susceptible and part of the general population that the AEGL values are intended to protect. These categories include children and infants, the elderly, asthmatics, pregnant women and the fetus, and individuals with pre-existing illnesses, diseases or metabolic disorders who would not ordinarily be considered in a severe or critical medical condition. Examples of susceptible subpopulations based on pre-existing illnesses include those with compromised pulmonary function (e.g., pneumoconiosis, emphysema, respiratory infections, smoking, immunologic sensitization due to prior exposures, and cystic fibrosis), hepatic function (e.g., alcoholism, hepatitis, and prior chemical exposures), cardiac function (e.g., dysrhythmias and coronary heart disease), and impaired renal or immunologic function (e.g., AIDS and systemic lupus erythematosus)

Hypersusceptible subpopulations are considered to comprise those individuals whose reactions to chemical exposure are unique and idiosyncratic; lie outside the range of distributions expected for the general population, including susceptible subpopulations; and constitute a relatively small component of the general population. For example, the AEGLs are intended to be protective of individuals with mild-to-moderate asthma but are not necessarily protective of those with severe asthma. Additionally, there are some asthmatics who, at any given time, could be suffering coincidentally acute asthmatic episodes at the time of a chemical emergency. Such subpopulations may be considered to comprise transient hypersusceptible individuals and would not necessarily be protected by the AEGLs. Examples of hypersusceptible subpopulations might include those with severely debilitating pulmonary, hepatic, or renal disorders or diseases, the elderly with serious debilities of primary physiologic systems, and those individuals with unique hypersensitivies (i.e., severe immune-type responses) to specific chemicals (e.g., 4,4'-methylene bis(2-chloroaniline); MOCA) or chemical classes (e.g., isocyanates). It is acknowledged that the AEGL values might not be protective under such circumstances.

Certain otherwise healthy individuals in the general population also may suffer transient periods of hypersusceptibility as a result of severe (reversible) short-term illnesses. For example, during recovery from a severe episode of acute upper respiratory infection, many nonasthmatic individuals can experience several weeks or more of bronchiolar hyperreactivity and bronchospasm following nonspecific exposure to respiratory irritants. This reversible condition is considered an example of transient hypersusceptibility, and it is acknowledged that the AEGL values may not be protective of individuals in such circumstances.

The nature of the dose-response relationships for immune-type responses among sensitive and hypersensitive individuals is highly complex and not well-understood. Additionally, in almost all instances, there is no clear line of demarcation that distinguishes susceptible from hypersusceptible individuals with respect to chemical responses that are not immune mediated, and there is no generic or medical guidance that can be followed for a wide range of chemical exposures. However, since most biologic responses are chemical-specific and are dependent on the mode of action of the substance in question, the issue of identifying and protecting groups of susceptible subpopulations is addressed by the NAC/AEGL Committee on a chemical-by-chemical basis. The committee uses all available data on the properties of the chemical and their relationship to both normal and compromised biochemical, physiologic, and anatomical systems in humans to identify and protect susceptible subpopulations. In the absence of data on the chemical in question, the use of structurally related chemicals and scientific judgment may be used to select UFs that provide protection for the public health.

2.5.3.3.4 Estimating the Range of Variability in a Human Population

The NAC/AEGL Committee estimates the range in variability of response to specific chemical exposures primarily on the basis of quantitative human data. Acceptable experimental data are more likely to be available for AEGL-1 and AEGL-2 endpoints than for AEGL-3 endpoints. For example, numerous studies have considered induction of bronchospasm after controlled exposures to sulfur dioxide (SO_2) in asthmatic and nonasthmatic individuals (see references below). There is marked individual variability in the severity of reaction to inhalation of low concentrations of SO_2. Asthmatics, individuals with hyper-reactive airways, smokers, and those with chronic respiratory or cardiac disease respond at relatively lower concentrations (Aleksieva 1983; Simon 1986). Susceptibility may also be increased in people over 60 years of age, but reports have not been consistent (Rondinelli et al. 1987; Koenig et al. 1993). By contrast, comparable human data for AEGL-3 tier concentrations are limited to anecdotal case reports.

Deliberations on phosphine AEGL development identified the possibility that children are more susceptible to phosphine exposure. This condition was suggested by two case reports describing the deaths of children, but not adults, after "comparable" phosphine exposures. However, both the children and the adults in question were present in somewhat restricted environments, suggesting comparable exposure levels. Based on these case reports, the NAC/AEGL Committee concluded that children may be more susceptible to phosphine exposure and selected UFs that would provide additional protection for children.

In cases in which quantitative human data are lacking for specific chemicals, but adequate data can be found for structurally or mechanistically similar agents, UFs may be selected by analogy to structurally similar chemicals and/or mechanism of action. For example, asthmatics are particularly susceptible to SO_2. Declines of more than 20% in FEV_1 have been documented after inhalation of 0.4-1 ppm for 2-15 min. The effects of SO_2 exposure are enhanced in normal and asthmatic individuals by moderate exertion (ventilation >40 liters per minute with mouth breathing), hyperventilation, and use of oral airways (Frank 1980; Koenig et al. 1981, 1982; Roger et al. 1985; Balmes et al. 1987; Linn et al. 1987; Horstman et al. 1988). Duration of bronchospasm is generally short, and these patients may develop tolerance with prolonged or repeated SO_2 exposure. These studies suggest that mouth-breathing asthmatics exposed to SO_2 develop bronchospasm at levels of approximately 33% of comparably exposed nonasthmatics. However, Schlesinger and Jaspers (1997) reported approximately 10-fold difference in susceptibility to SO_2. Schlesinger (1999) also reported 5-fold difference in susceptibility for nitrogen oxide (NO_2). Therefore, a default UF of 10 is generally used to account for the differences in the potential broad range of human susceptibility to respiratory irritants. The NAC/AEGL Committee is aware that the variation in responses of asthmatics to respiratory irritants may range from mild to severe. A UF of less than 10 might be used when scientific evidence shows that a smaller UF will be protective of health.

Children and infants are often considered to be a susceptible subpopulation. There is a general belief that children and infants are more susceptible to the effects of toxic substances than adults. Much of this belief is predicated upon the fact that children and particularly infants possess immature or developing biochemical, physiologic, and anatomical systems that are not adequate to combat the adverse affects of toxic chemicals. Further, it is believed that in certain instances the toxic effects of chemicals may permanently damage or alter the growth and function of developing organs and organ system. The potential for greater susceptibility to chemical substances by children and infants has been reviewed by the NRC (1993b). The report indicates that there are limited data on the relative toxicity of pesticides and other xenobiotic compounds in immature and mature humans. Consequently, the NRC (1993b) focused on laboratory animal studies, age-related pharmacokinetic and pharmacodynamic differences, and pharmacologic data from controlled clinical investigations with humans. The NRC (1993b) concluded that the mode of action is generally similar in mammalian species and across age and developmental stages within species. The NRC also concluded that children may be more or less susceptible than adults to pesticide toxicity, depending on the chemical, but that the quantitative differences in toxicity between the age groups are usually less than a factor of approximately 10.

Although many reports have been published on the pharmacokinetic differences of pharmacologic agents and other chemicals in children and adults, the data cannot be translated into meaningful dose-response relationships to make valid quantitative comparisons in the absence of specific biologically relevant endpoints. Bruckner and Weil (1999) summarized the biologic factors that may influence the responses of adolescents to chemical exposure. Based on the limited data available, the extent to which significant differences exist in the susceptibility of children and infants and of adults is largely unknown. However, the difference is generally considered to be within a factor of 10 (NRC 1993b), most of the differences in susceptibility being on the order of 2- to 3-fold. It is highly probable that any differences are chemical-specific and also related to specific developmental stages of children and infants. Within the context of the AEGL program, this issue is further complicated by the consideration of once-in-a-lifetime inhalation exposures of less than 1-8 h. The discussion at the beginning of this section indicates that there is a paucity of data on age-related differences and the young can be more or less susceptible than adults to xenobiotic exposures, depending upon the chemical or chemical class in question. However, it is believed that UFs applicable to other susceptible subpopulations are adequate to protect children and infants when decisions are based on the weight of the evidence on a chemical-specific basis. It is important that all of the relevant information on the chemical be considered when making judgments about selection of the appropriate UF for all factors that contribute to differences in susceptibility.

In summary, the maximum variation in responses of susceptible subpopulations are believed to generally range between 3- and 10-fold of the responses for healthy individuals. All information on the chemical, including its mechanism of action, the biologic responses, physical and chemical properties, data on structurally related chemical analogues, and toxic endpoints in question, is considered and informed professional judgment is used when determining appropriate UFs. Information about similarities and differences in toxicokinetics and toxicodynamics is used when available to modify the UF used. Therefore, in the absence of the type of information described above, a default UF of 10 is used in the development of AEGLs to account for susceptible human subpopulations. If such data or information are available, the NAC/AEGL Committee may conclude that a UF of 3 or 1 will provide adequate protection. The final selection of UFs is based on a weight-of-evidence decision that considers the chemical's mechanism of action, the available human and animal data, and the susceptible subpopulations that may be at risk.

2.5.3.4 Intraspecies UFs—NAC/AEGL Guidelines

Guidelines followed by the NAC/AEGL Committee to select UFs are presented below. In each section, there is a list of questions that should be addressed to support the rationale for the choice of the UF used. The guidelines are organized into categories for convenience. However, more than one guideline may be applied to the selection of any one UF.

In general, in the absence of data or information to the contrary, the default value for the intraspecies UF is 10. However, a UF of 3, or even 1, may be used if credible information or data are available. The UF is determined on a case-by-case basis and may be dependent on the information or data available on humans or animals; the specific biologic, mechanistic, and physical and chemical properties of the chemical; and the health-effect endpoint in question. The following cases provide general guidance for the most common circumstances encountered by the NAC/AEGL Committee in selecting UFs.

2.5.3.4.1 Toxic Effect Is Less Severe Than Defined for the AEGL Tier

If the toxicologic effects described in the chosen database are judged to be somewhat less severe than those defined for the AEGL tier in question, an intraspecies UF less than 10-fold may be used.

The rationale for the selection of this UF should include the following:

1. Description of the toxicologic endpoint of concern selected and how it relates to the AEGL severity tier in question.
2. Comment on the slope of the dose-response relationship if possible and explain how this impacts the UF.

2.5.3.4.2 Susceptible Individuals Used

If individuals representative of a susceptible subpopulation are used as subjects in controlled humans studies, and the AEGL is to be calculated based on effects observed in those individuals, an intraspecies UF of less than 10-fold may be used.

The rationale for the selection of this UF should include the following:

1. Description of the condition that made the individual susceptible.
2. Description of the adverse effects assessed.
3. Discussion of the range of response of the humans tested.
4. Description of the differences or similarities in the responses of susceptible individuals qualitatively and quantitatively as compared with nonsusceptible individuals.

2.5.3.4.3 Age, Life Stage, and Physical Condition Differences

When available data indicate that certain groups based on age, life stage, or physical condition may be uniquely susceptible in contrast to the general population, an intraspecies UF of 10 is generally used.

The rationale for the selection of this UF should include the following:

1. Description of the toxicologic endpoints that differ between humans of different age groups.
2. Discussion of the magnitude of a difference. For example, quantitatively, how much does the response differ, or what qualitative information indicates differences among age groups?

2.5.3.4.4 Response of Normal and Susceptible Individuals to Chemical Exposure is Unlikely to Differ for Mechanistic Reasons

In those cases in which the mode or mechanism of action is such that the response elicited by exposure to the chemical by different subpopulations is unlikely to differ, an intraspecies UF of 3-fold is generally used. Typically, this response involves a direct-acting mechanism of toxicity in which metabolic or physiologic differences are unlikely to play a major role. A steep dose-response curve may also be an indication of little variation within a population and is factored into the weight-of-evidence considerations for UF determination.

The rationale for the selection of this UF should include the following:

1. Description of the mode of action.
2. Discussion of why the response to chemical exposure is unlikely to differ between healthy and susceptible individuals and whether metabolic or physiologic differences are likely to be an issue.

2.5.3.4.5 Mode or Mechanism of Action Is Unknown

When the mode or mechanism of toxic action is uncertain, or unknown, or metabolic factors may play an important role, and/or when a broad range of responses to chemical exposure is observed, there is concern that there may be large differences in susceptibility between individuals. In those cases, an intraspecies UF of 10 may be applied.

The rationale for the selection of this UF should include the following:

1. Description of the toxicity reported and the uncertainty associated with the chemical's mechanism of action or other factors.
2. Statement on why the effects seen add uncertainty to the assessment.

2.5.3.4.6 UFs That Result in AEGL Values That Conflict with Actual Human Exposure Data

When AEGL values are initially derived, the candidate range of values is compared with the known spectrum of supporting data on the chemical. In a weight-of-the-evidence approach, conflicts between the candidate AEGLs (generally derived from animal data) and the supporting data (either animal data or human data) may lead to the conclusion that the UFs utilized in the calculations are inappropriate, because they conflict with other specific and highly relevant human data. In that case, the candidate AEGLs are revised to reflect the supporting data. In other cases in which the AEGL may conflict with an existing standard or guideline, the comparative basis of the two values may be evaluated to see if the discrepancy is justified or resolvable.

The rationale for the selection of this UF should include the following:

1. A statement on why the use of UFs initially selected conflicts with the published evidence.

2.6 GUIDELINES AND CRITERIA FOR SELECTION OF MODIFYING FACTORS

2.6.1 Definition

In addition to the UFs discussed above, an additional modifying factor may be necessary when an incomplete database exists. Hence, the modifying factors represent an adjustment for uncertainties in the overall database or for known differences in toxicity among structurally similar chemicals. The modifying factor "reflects professional judgment on the entire data base available for the specific agent" and is applied on a case-by-case basis (NRC 1993a, p. 88). The modifying factor may range from 1- to 10-fold. The default value is 1.

2.6.2 Use of Modifying Factors to Date in the Preparation of AEGL Values

Modifying factors have been used in AEGL documents for four chemicals recently published by the NRC (2000b). Modifying factors of 2 or 3 are under consideration for chemicals currently undergoing review to account for (1) a limited data set, (2) instances in which the adverse effects used to set the AEGL value are more severe than those described in the AEGL definition, and (3) the differential toxicity of chemical isomers.

2.7 GUIDELINES AND CRITERIA FOR TIME SCALING

AEGLs are derived for 30-min, 1-h, 4-h, and 8-h exposure durations to meet a wide range of needs for government and private organizations. AEGLs for 10-min exposure durations will be developed for all future chemicals addressed by the NAC/AEGL Committee, and 10-min AEGLs will be developed for the first four chemicals published by the NRC (2000). Experimental animal and controlled human exposure-response data and data from human exposure incidents often involve exposure durations differing from those specified for AEGLs. Therefore, AEGL development usually requires extrapolation from the reported exposure duration and chemical concentration of a toxic endpoint to an equivalent concentration for an AEGL-specified exposure period. This section discusses that concept, the published scientific literature, the methodologies used for extrapolation, and examples of the application of these methodogies to specific chemicals for the development of AEGL values.

2.7.1 Overview

The NRC (1993a) guidelines for developing short-term exposure levels address extrapolation of the effects of genotoxic carcinogens from long-term to short-term exposures. Only limited NRC guidance is provided for approaches or methodologies for the extrapolation of reported acutely toxic effects to shorter or longer durations of exposure.

The relationship between dose and time to response for any given chemical is a function of the physical and chemical properties of the substance and the unique toxicologic and pharmacologic properties of the individual substance. Historically, the relationship according to Haber (1924), commonly called Haber's law (NRC 1993a) or Haber's rule (i.e., $C \times t = k$, where C = exposure concentration, t = exposure duration, and k = a constant) has been used to relate exposure concentration and duration to effect (Rinehart and Hatch 1964). This concept states that exposure concentration and exposure duration may be reciprocally adjusted to maintain a cumulative exposure constant (k) and that this cumulative exposure constant will always reflect a specific quantitative and qualitative response. This inverse relationship of concentration and time may be valid when the response to a chemical is equally dependent upon the concentration and the exposure duration. However, an assessment by ten Berge et al. (1986) of LC_{50} data for certain chemicals revealed chemical-specific relationships between exposure concentration and exposure duration that were often exponential. This relationship can be expressed by the equation $C^n \times t = k$, where n represents a chemical-specific, and even a toxic endpoint-specific, exponent. The relationship described by this equation is basically in the form of a linear regression analysis of the log-log transformation of a plot of C vs t (see Section 2.7.5.3). ten Berge et al. (1986) examined the airborne concentration (C) and short-term exposure duration (t) relationship relative to death for approximately 20 structurally diverse chemicals and found that the empirically derived value of n ranged from 0.8 to 3.5 among this group of chemicals (see Table 2-1). Hence, these workers showed that the value of the exponent (n) in the equation $C^n \times t = k$ quantitatively defines the relationship between exposure concentration and exposure duration for a given chemical and for a specific health-effect endpoint. Haber's rule is the special case where n = 1. As the value of n increases, the plot of concentration vs time yields a progressive decrease in the slope of the curve.

In cases in which adequate data are available, the NAC/AEGL Committee conducts an analysis of chemical-specific toxicity and exposure data to derive a chemical-specific and health-effect-specific exponent (n) for use in extrapolating available exposure data to AEGL-specified exposure durations. If data are not available for empirically deriving the exponent n, the NAC/AEGL Committee identifies the most appropriate value for n by comparing the resultant AEGL values derived using n = 1 and n = 3. The value of n = 1 has been used historically by others and results in rapid reductions in concentrations when extrapolated to longer exposure periods and rapid increases in concentrations when extrapolated to shorter exposure periods. Based on the work of ten Berge et al. (1986), n = 1 represents the estimate of the lower boundary of the value of n. The value of n = 3, an estimate of the upper boundary of the value of n (ten Berge et al. 1986), results in less rapid rates of decrease in estimated effect concentrations when extrapolated to longer exposure periods and to less rapid rates of increase in estimated effect concentrations when extrapolated to shorter exposure periods. This range of values of n from 1 to 3 encompasses approximately 90% of the chemicals examined

TABLE 2-1 Values of n from ten Berge et al. (1986)

	Value of n (average)
Systemic Chemicals	2.7
Hydrogen Sulfide	2.2
Methyl *t*-butyl ether	2
Methylenechlorobromide	1.6
Ethylenedibromide	1.2
Tetrachloroethylene	2
Trichloroethylene	0.8
Carbon tetrachloride	2.8
Acrylonitrile	1.1
Irritants	
Ammonia	2
Hydrogen Chloride	1
Chlorine pentafluoride	2
Nitrogen dioxide	3.5
Chlorine	3.5
Perfluoroisobutylene	1.2
Crotonaldehyde	1.2
Hydrogen Fluoride	2
Ethylene imine	1.1
Bromine	2.2
Dibutylhexamethylenediamine	1

Range of n	No. of Chemicals per Range	Cumulative No. of Chemicals
0.8-1.5	8	8
1.51-2.0	6	14
2.01-2.5	2	16
2.51-3.0	2	18
3.01-3.5	2	20

by ten Berge et al. (1986). In selecting a value for n when the derivation of n is not possible, the NAC/AEGL Committee evaluates the resultant AEGL values determined with either the upper or the lower boundary value of n (1 or 3) within the context of other supporting data to determine the reasonableness of the extrapolated AEGL value. A value of n = 1 is used when extrapolating from shorter to longer exposure durations and a value of n = 3 is used when extrapolating from longer to shorter durations. The resultant AEGL value is then compared with supporting data to determine the scientific reasonableness of the derived AEGL value and the process favors the use of a value for n that results in an AEGL value that best fits the supporting data.

In summary, since toxicity data are often not available for any or all of the AEGL specified time periods, temporal extrapolation is usually necessary to generate scientifically credible values for the AEGL time points. However, it is important to point out that the relevant data, together with scientific judgment, are used to determine the extent of temporal extrapolation and its validity in AEGL derivations. This is underscored by the fact that errors in the estimated exposure concentration–exposure duration relationship (i.e., the value of n) can progressively increase the magnitude of the uncertainty of the derived AEGL value as the time from the empirical data point to the extrapolated data point increases.

Therefore, extrapolation of 10-min exposure data to a 4-h or 8-h AEGL value requires more supporting data and/or assumptions than the extrapolation of 10-min exposure data to a 30-min or 1-h AEGL. Additionally, extrapolation of 4-h or 8-h exposure data to a 10-min AEGL value also requires more supporting data than extrapolation of 8-h exposure data to a 1-h or 30-min AEGL. As a result of the potential uncertainties in dose-response relationships associated with exposure durations of less than 10-min (a transition period in breathing rate, physiologic responses, scrubbing efficiency/saturation, chemical-specific issues, and other factors) and in the absence of definitive supporting data, extrapolations using 4-h and 8-h empirical data generally are not extended to 10-min AEGL values but are assigned the same value as the 30-min AEGL. Finally, because of the uncertainties for very short exposure durations, no AEGL values for exposure periods of less 10-min are currently recommended.

2.7.2 Summary of Key Publications on Time Scaling

Several investigators have studied the relationship of exposure duration and exposure concentration as related to the toxic response to airborne chemicals (Flury 1921; Haber 1924; Rinehart and Hatch 1964; ten Berge et al. 1986; ECETOC 1991; and Pieters and Kramer 1994).

On the basis of observations and studies with gases such as phosgene, Haber (1924) found that for certain chemicals the product of the exposure duration multiplied by the exposure concentration was constant for a specific response or toxic endpoint (i.e., lethality). In experiments with cats, Haber found that the product of a specific concentration and exposure time would result in 100% lethal response and that as long as this product value was maintained, regardless of the specific exposure concentration or duration, the response was consistent. This linear relationship became known as Haber's rule. Similarly, Flury (1921) found that inhalation of phosgene exhibited a linear relationship, $C \times t = E$, where E represents the onset of pulmonary edema. Obviously, the cumulative exposure constant may relate to any number of responses or physiologic endpoints. However, the information reported by Haber is limited to a small number of chemicals or chemical classes, and substantial quantitative data derived from controlled studies are lacking.

Historically, Haber's rule has been used for time concentration extrapolations (EPA 1994). This relationship assumes that each unit of damage is irreversible, that no repair takes place during the exposure period and, therefore, that each unit of exposure is 100% cumulative. However, this is generally not the case for acutely toxic responses to short-term exposures. The relationship between concentration and duration of exposure as related to lethality was examined by ten Berge et al. (1986) for approximately 20 irritant or systemically acting vapors and gases. The authors subjected the entire individual animal data set to probit analyses with exposure duration and exposure concentration as independent variables. They used the method of Finney (1971) to investigate the fit of the data into a probit model on the basis of a maximum likelihood estimate. In re-evaluating the raw data for these chemicals, it was found that the linear relationship described by Haber's rule, $C \times t = k$ was not always a valid predictor of lethality. An exponential function ($C^n \times t = k$), where the value of n ranged from 0.8 to 3.5 for different chemicals, was a more accurate quantitative descriptor. These authors derived empirically based, chemical-specific regression coefficients for exposure duration and exposure concentration, as well as chemical-specific values for n. The values for n for the 20 chemicals studied ranged from 0.8 to 3.5. The analyses indicated that the concentration-duration relationship for lethality was described more accurately by the exponential function ($C^n \times t = k$) and that Haber's rule was appropriate for only a limited number of chemicals. On the basis of the results of the analyses, ten Berge et al. (1986) concluded that the concentration-time relationship (i.e., value for n) should be determined empirically from acute inhalation exposure toxicity data on a chemical-specific basis.

2.7.3 Summary of the Approaches That May Be Taken for Time Scaling

A tiered approach to generating toxicity values for time scaling is taken by the NAC/AEGL Committee to derive AEGL values from empirical data.

This approach is summarized below. Each of the approaches and the circumstances under which they are, or could be, used are discussed subsequently in this section.

1. If appropriate toxicologic data for the exposure concentration–exposure duration relationship of a specific health-effect endpoint are available for the AEGL-specified exposure periods, use the empirical data directly.
2. If empirical exposure concentration–exposure duration relationship data are available, albeit they do not coincide with AEGL-specified exposure periods, use the available data to derive values of n and extrapolate the AEGL values using the equation $C^n \times t = k$. If supporting data are inconsistent with the extrapolated AEGL value, the value of n might be modified to reconcile the difference. If definitive supporting data for 10-min exposures are not available when extrapolating from 8-h empirical data, the 10-min AEGL generally is assigned the same value as that extrapolated for the 30-min AEGL.
3. If no empirical exposure concentration–exposure duration relationship data are available to derive a value of n, a value of n = 1 for extrapolating from shorter to longer exposure durations and a value of n = 3 for extrapolating from longer to shorter exposure durations should be tested initially. The scientific reasonableness of the selection of the estimated lower and upper boundaries of n (n = 1 and n = 3) is then evaluated by comparing the resultant AEGL values with all other supporting data. If appropriate, the final value(s) of n may be modified to reconcile differences between extrapolated AEGL values and the supporting data. If definitive supporting data for 10-min exposures are not available when extrapolating from 8-h empirical data, the 10-min AEGL generally is assigned the same value as that extrapolated for the 30-min AEGL.
4. If there are no supporting data to evaluate selected values of n, a default value of n = 1 for extrapolating from shorter to longer exposure periods and a default value of n = 3 for extrapolating from longer to shorter exposure periods should be selected. If definitive supporting data for 10-min exposures are not available when extrapolating from 8-h empirical data, the 10-min AEGL generally is assigned the same value as that extrapolated for the 30-min AEGL. In the absence of other data, the resultant AEGL values are thought to be protective and scientifically credible.

The remainder of this section of the guidance provides more detailed information on the approaches stated above.

2.7.4 Use of Empirical Data Available for AEGL-Specified Exposure Durations

If toxicity data are available for all four AEGL-specified exposure periods, there is no need to derive values of n, and the empirical data for each exposure period can be used directly. However, it is rare that toxicity data are sufficiently comprehensive to encompass all the AEGL-specified exposure periods from 10 min to 8 h. Further, there are instances in which empirical data are not available to estimate n and predict the exposure concentration–exposure duration relationship using $C^n \times t = k$. Therefore, the sequential approaches used by, or available to, the NAC/AEGL Committee to establish AEGL values for the specified exposure periods are discussed in the following sections.

2.7.5 Derivation of Values of n When Adequate Empirical Data Are Available for Durations Other Than the AEGL-Specified Exposure Durations

A key element in the procedure of time scaling is the use of a value or values for n in the equation $C^n \times t = k$. If empirical data for exposure durations other than the AEGL-specified exposure periods are available to quantify the exposure concentration–exposure duration relationships for a health-effect endpoint, including lethality, the value of n should be derived using the method of calculation described in this section. It is believed that empirically derived values of n are scientifically more credible than a default value of n = 1 (Haber's rule) or attempting to derive an alternate value of n.

2.7.5.1 Selection of Appropriate Health-Effect Endpoint for Deriving a Value for n

The first step in any time scaling methodology is the selection of the health-effect endpoint of concern. Clearly, the health-effect endpoint selected should be consistent with the definition of the AEGL tier being determined. Further, the endpoint should be unambiguous and consistently observed at all reported exposure durations. For example, death is an unambiguous endpoint and a quantitatively determined index of toxicity; the LC_{50} is a response rate that can be compared reliably among exposures at different time periods. The use of the LC_{50} as an index of toxicity is ideal, because it is a statistically derived concentration that is not subject to the vagaries of dose selection and exhibits less variability in response than any other experimental endpoint.

DERIVATION OF AEGL VALUES   99

Death is included in the AEGL-3 definition and is used for estimating the value of n.

A comparable endpoint for the AEGL-1 and AEGL-2 tiers would be an ED_{50} (the dose that causes a specific response in 50% of the test subjects) for a precisely defined toxic or health-effect endpoint that is consistent with the definition of the AEGL tier in question. The actual endpoint is often difficult to determine in most experiments, because the observed effects are often a continuum from mild to severe and generally not reported in a format precise enough to determine a reliable ED_{50} value. Further, incidence data for non-lethal effects are not always reported. For these reasons, the concentration-response relationship and the value of n derived from lethality data have often been applied to both the AEGL-3 and the AEGL-2 exposure-period extrapolations. However, in instances in which the mechanism of toxicity causing the health effect of concern at the AEGL-2 tier is thought to be different from that causing lethality, the value of n derived from LC_{50} data should not be used. Under these circumstances, AEGL-2 values can be developed by selecting the upper and lower boundaries of n (n = 3 and n = 1) for extrapolation from longer to shorter and shorter to longer exposure periods, respectively. The resultant AEGL-2 values should be evaluated within the context of other supporting data to evaluate the reasonableness of the values of n selected. In the absence of supporting data, the AEGL values determined using n = 3 and n = 1 should be utilized as described in No. 3 in Section 2.7.3.

Selection of appropriate endpoints for AEGL-1 values per se represents a unique and often difficult task. Based on its experience to date, the NAC/AEGL Committee has found no rigorous data available from which values of n could be derived for AEGL-1 type of endpoints for any chemical. The derivation of AEGL-1 values is discussed later in this section.

2.7.5.2 Criteria for Adequate Empirical Data for Deriving Values of n

After determining the health-effect endpoint to be used in deriving the value(s) for n, the next step is to evaluate the quality and the quantity of the data to be used in the derivation. Obviously, two data points will define the slope of a curve describing the exposure concentration-duration relationship. However, the validity and, hence, the values of n will depend on many factors, including the scientific soundness of the exposure concentration-duration data, the length of the empirical exposure duration(s) relative to the AEGL-specified exposure periods, and the known or perceived similarities in effects and mechanism of action of the chemical at the reported exposure concentrations and durations. Generally, three empirical data points will improve the scientific validity of the slope and the estimated values for n, and the validity is likely to increase with an increase in the number of empirical data points used to derive n, provided that there is a reasonable fit of these data points.

2.7.5.3 Curve Fitting and Statistical Testing of the Generated Curve

Once the health-effect endpoint and data points describing the exposure concentration-duration relationship have been selected, the values are plotted and fit to a mathematical equation from which the AEGL values are developed. There may be issues regarding the placement of the exponential function in the equation describing the concentration-duration relationship (e.g., $C^n \times t = k$ vs $C \times t^m = k_2$ vs $C^x \times t^y = k_3$). It is clear that the exposure concentration-duration relationship for a given chemical is directly related to its pharmacokinetic and pharmacodynamic properties. Hence, the use and proper placement of an exponent or exponents to describe these properties quantitatively is highly complex and not completely understood for all materials of concern.

The quantitative description of actual empirical data of the concentration-duration relationship can be expressed by any of a number of linear regression equations. In the assessment of empirical data reported by ten Berge et al. (1986), these workers quantified the exposure concentration-duration relationship by varying the concentration to the nth power. Since raising c or t or both to a power can be used to define quantitatively the same relationship or slope of the curve and to be consistent with data and information presented in the peer-reviewed scientific literature, the equation $C^n \times t = k$ is used for extrapolation. It must be emphasized that the relationship between C and t is an empirical fit of the log transformed data to a line. No conclusions about specific biologic mechanisms of action can be drawn from this relationship.

The preferred approach is to use a statistical methodology, which utilizes all the individual animal data and generates a maximum likelihood estimate with 95% confidence limits. When individual animal data are available, the method of Finney (1971) is preferred. This method has been incorporated into a computer program (provided to the NAC/AEGL Committee by Dr. ten Berge from The Netherlands).

Unfortunately, the individual animal data are often unavailable and only LC_{50} values are listed. In this case, a linear regression analysis of the log-log transformation of the concentration-duration data is performed as described below:

When concentration-duration data are plotted on a log-log plot, they generally fall along a straight line. For that reason, a simple linear regression

(Alder and Roessler 1968) is run on the data to generate the mathematical curve. The basic linear regression equation is in the form

$$Y = a + bX,$$

where Y is the predicted value of the dependent variable, X is the value of the independent variable, a is the Y intercept and b is the slope of the line.

This is the form of the log-transformation of the nonlinear $C^n \times t = k$ equation to a linear equation (see below):

$$\log C = (\log k)/n + (-1/n) \times \log t,$$

where C is the predicted value of the concentration to cause an effect at exposure duration t. The $(\log k)/n$ is the Y intercept of the plot of log C against log T, and $-1/n$ is the slope of the plot of log C against log T.

$$C^n \times t = k$$
$$\log(C^n \times t) = \log k$$
$$n \times \log C + \log t = \log k$$
$$n \times \log C = \log k - \log t$$
$$\log C = (\log k)/n - (\log t)/n$$
$$\log C = (\log k)/n - (1/n) \times \log t$$

The regression coefficient or slope, b, returns the slope of the linear regression line through data points X and Y. The slope (rate of change along the regression line) is the distance between the Y values of the two points divided by the distance between their respective X values. The regression coefficient is calculated as

$$b = \frac{N\Sigma XY - (\Sigma X)(\Sigma Y)}{N\Sigma X^2 - (\Sigma X)^2},$$

where N = the number of observations,

or

$$-1/n = \frac{N\Sigma(\log t)(\log C) - (\Sigma \log t)(\Sigma \log C)}{N\Sigma(\log t)^2 - (\Sigma \log t)^2}.$$

The above is solved for n.

The validity of the derived values of n is dependent on the degree of correlation among the various concentration-duration data points used to construct the curve and the equation. Normally a coefficient of determination (r^2) is calculated as a measure of how well the generated curve (linear in this case) fits the data points. If $r^2 = 0$, then the data do not fit a linear relationship. If $r^2 = 1$, then the data exhibit a strong linear relationship. If the number of data points are 3 and the real value of $r = 0$, "the chance of obtaining a fairly high correlation coefficient for the sample is greater than the chance of obtaining a small correlation coefficient" (Alder and Roessler 1968). If the number of data points are 4, "the chance of obtaining a particular correlation coefficient is equal to that of obtaining any other" (Alder and Roessler 1968). Since the number of data points typically available are only in the range of 3 or 4 values, the use of r^2 to measure how well the data fit the generated curve is not meaningful. Therefore, informed professional judgment is exercised by the NAC/AEGL Committee.

Given the fact that the distribution of r for low numbers of observations (typically 3 or 4 data points for time scaling) cannot be fit to a normal curve, meaningful statistical tests of the fit of the regression line (used to derive n) to the data cannot be performed. Even with those shortcomings, a regression analysis of the data as previously described gives the best fit of a line to the data. An inspection of the regression line vs the data also will show the reasonableness of the fit and, hence, the reasonableness of the derived value for n. This approach is generally the best when empirical data are used to derive n values for developing AEGL values for specified exposure durations. As stated earlier, it must be emphasized that when deriving or selecting a value for n, the NAC/AEGL Committee evaluates the resultant AEGL values within the context of other supporting data to determine the reasonableness of the extrapolated values. This is done even when the value of n is derived from empirical data that describe the exposure concentration-duration relationship. The NAC/AEGL Committee uses a value for n that results in AEGL values that best fit the supporting data. Therefore, there is no substitute for informed professional judgment based on careful review, evaluation, and discussion of all available data.

2.7.5.4 Examples of NAC/AEGL Committee Derivations of Values of n from Empirical Data

During the course of AEGL development, the NAC/AEGL Committee has used empirically based derivations of n in the equation $C^n \times t = k$ for time scaling to AEGL-specified exposure periods. Guidelines have been developed from this experience and are presented in the final part of this section.

2.7.6 Selection of Values of n When Adequate Empirical Data Are Not Available to Derive Values for n

When adequate data describing exposure concentration-duration relationships for a specific chemical and toxic endpoint of interest are not available, an alternative approach to estimating this relationship quantitatively must be followed. The approach used by the NAC/AEGL Committee involves the application of the equation $C^n \times t = k$ and the selection of a value or values of n that results in AEGL values that best fit the supporting data for the chemical and toxic endpoint in question. It is important to distinguish the difference between the *derivation* of values of n as described in the preceding section and the *selection* of values of n as described in this section.

An evaluation of the analysis of values of n by ten Berge et al. (1986) served as the basis to select the limits used by the NAC/AEGL Committee.

The lowest value of n was 0.8 and the highest value of n was 3.5. Approximately 90% of the values of n range between $n = 1$ and $n = 3$. Consequently, these values were selected as the reasonable lower and upper bounds of n.

In the absence of data to derive a value for n, the NAC/AEGL Committee selects values for n of 1 and 3, depending on an extrapolation from shorter to longer durations or longer to shorter durations, respectively. The value of n is then used in the equation $C^n \times t = k$ to extrapolate from empirically reported exposure concentrations and exposure durations to the AEGL-specified exposure durations. The committee then selects the derived AEGL values in accordance with the supporting data.

2.7.6.1 Selection of Values of n When Extrapolating from Shorter to Longer Exposure Durations

As discussed previously, a value of $n = 1$ represents the lower range of the exposure concentration–exposure duration relationship. If the exponent $n = 1$ is used in the equation $C^n \times t = k$, there is a rapid decrease in extrapolated values when extrapolations are made from shorter to longer exposure periods (see Figure 2-2). The extrapolated values are lower and, hence, represent a conservative estimate of the AEGL value. A value of $n = 3$ represents a value in the upper range for the exposure concentration-duration relationship and results in a less rapid rate of decrease when extrapolating from shorter to longer exposure periods. Therefore, the extrapolated AEGL values for longer exposure periods are higher and, hence, less conservative in terms of protecting human health (see Figure 2-2).

When empirical data are not available for deriving a value of n, the

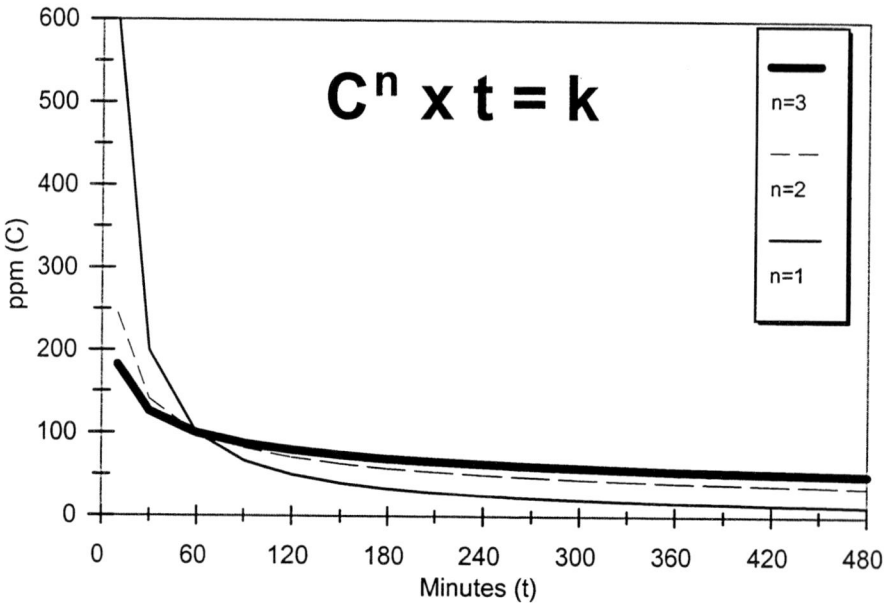

FIGURE 2-2 Effects of varying n in the equation $C^n \times t = k$. *Short- to Long-Duration Extrapolations:* Note that when extrapolating from 60 min to longer exposure durations, the lower the value of n, the lower the extrapolated value. Therefore, when extrapolating from short- to long-exposure durations, a value of $n = 1$ yields a more conservative value than any value of n that is >1. *Long- to Short-Duration Extrapolations:* Conversely, when extrapolating from 60 min to shorter exposure durations, the higher the value of n, the lower the extrapolated value. Therefore, when extrapolating from long to short exposure durations, a value of $n = 3$ yields a more conservative value than any value of n that is <3.

NAC/AEGL Committee develops tentative AEGL values from shorter to longer exposure durations using $n = 1$ in the equation $C^n \times t = k$ and evaluates these values with all other supporting data to determine their scientific reasonableness. Therefore, a weight-of-evidence test is applied to the tentative AEGLs by comparing these values to the supporting data to determine the most scientifically credible AEGL values. In instances in which supporting data indicate that the tentative AEGL developed using a value of $n = 1$ is inaccurate, the AEGL may be adjusted to scientifically accommodate the supporting data. If there are no supporting data indicating that the derived AEGL should be adjusted, a value of n = 1 is used to account for the uncertainty of the concentration-endpoint relationship at longer exposure durations.

2.7.6.2 Selection of Values of n When Extrapolating from Longer to Shorter Exposure Durations

When extrapolating from longer to shorter exposure durations using the equation $C^n \times t = k$ and a value of n = 1, there is a relatively rapid increase in the extrapolated values (see Figure 2-2). Under these circumstances, the derived AEGL value represents a relatively high estimate of the toxic endpoint concentration at shorter exposure durations and is, therefore, a less conservative value. When extrapolating from longer to shorter exposure durations using a value of n = 3, there is a less rapid rate of increase in the derived AEGL value. As a result, the extrapolated AEGL value is more conservative when selecting a value of n = 3 (see Figure 2-2).

In circumstances in which the NAC/AEGL Committee selects a value for n to derive AEGL values from empirical data for longer to shorter exposure periods, tentative AEGLs are derived using a value of n = 3 and then compared with the derived AEGL values with all other relevant data. Again, this represents a weight-of-evidence approach to selecting a value of n for the most scientifically credible AEGL values. In instances in which the supporting data indicate that the tentative AEGL developed using a value of n = 3 is too high or too low, the AEGL may be adjusted to scientifically account for the supporting data. If no supporting data indicate that the derived AEGL should be adjusted, a value of n = 3 should be used to accommodate for the uncertainty of the exposure concentration-duration relationship for the shorter exposure durations. If definitive supporting data for 10-min exposures are not available when extrapolating from 8-h empirical data, the 10-min AEGL generally is assigned the same value as that extrapolated for the 30-min AEGL.

2.7.7 Special Considerations in the Time Scaling of AEGL-1 and AEGL-2 Values

The previous descriptions of approaches to time scaling for toxic endpoint concentrations are most applicable to the derivation of AEGL-3 values, because unequivocal data relating the concentration required to cause an effect to the exposure duration are LC_{50} data. Lethality is an unambiguous endpoint that does not involve gradations of severity or incidence that are often difficult to quantify precisely (e.g., pulmonary congestion or edema or irritation in the respiratory tract involving variations in both degree and area affected). With respect to AEGL-2 values, it is far more difficult to quantify and achieve consensus on gradations in nonlethal toxic effects with respect to severity and incidence in a manner that readily results in a simple, quantitative endpoint exposure concentration-duration relationship. Further, the LC_{50} is a statistically derived value in the midpoint of the dose-response curve that is less subject to the vagaries in response at the extremes of the exposure regimen. For these reasons, the NAC/AEGL Committee primarily has used LC_{50} data in the derivation of exposure-duration scaling relationships. Subsequently, these quantitative relationships have been used to derive both AEGL-2 and AEGL-3 values and occasionally AEGL-1 values. This approach is believed to be scientifically credible if the mechanism of toxicity for AEGL-2 and AEGL-3 is known or thought to be similar.

It is recognized that the time-scaling relationship observed with a lethality AEGL-3 endpoint may not accurately describe the irreversible-effects or impairment-of-escape endpoint used for the AEGL-2. However, the NAC/AEGL Committee compares the AEGL-2 values against the supporting data to assess the reasonableness of the AEGL-2 determinations. Based on this assessment, adjustments are made to better fit the supporting data. If there are data that suggest different toxicologic mechanisms for lethal effects and AEGL-2 health effects, selected values of n should be used for the development of the AEGL values. The upper and lower bounds of $n = 3$ and $n = 1$ should be used for extrapolation from longer to shorter and from shorter to longer exposure periods, respectively. The resultant AEGL-2 values should then be evaluated using all supporting data and adjusted or maintained accordingly.

A difficult application of time scaling is encountered when attempting to derive AEGL-1 values. The AEGL-1 value defines the airborne concentration that distinguishes detection from discomfort. As a result, there is greater difficulty in attempting to quantify this often subjective level with respect to severity and incidence in a manner sufficient to derive an exposure concentration-duration relationship than to quantify the AEGL-2. This quantification is further complicated by the nature of the biologic endpoint that one is attempting to quantify. For example, the concentration for odor detection may actually decrease over time because of olfactory fatigue, as in the case of hydrogen sulfide. With respect to mild sensory effects, they are generally not cumulative over a range of exposures of 10 min to 8 h. Hence, the same AEGL-1 value may be assigned to all AEGL-specified exposure periods. In certain instances in which experimental data suggest that the sensory effects may increase because of the cumulative dose over time, the 10-min, 30-min,

DERIVATION OF AEGL VALUES   107

and 1-h values may be the same, but may be lower for the 4-h and 8-h AEGL exposure durations.

In the case of certain sensory irritants, the AEGL values may be constant across all AEGL time periods, because this endpoint is considered a threshold effect, and prolonged exposure will not result in an enhanced response. In fact, individuals may adapt or become inured to sensory irritation provoked by exposure to these chemicals over these exposure periods such that the warning properties are reduced.

2.7.8 Time Scaling—Guidelines for the NAC/AEGL Committee Approach

This section presents a compilation of time-scaling guidelines, which are used when deriving AEGL values for different time periods. As stated earlier, it must be emphasized that when deriving or selecting a value for n, the NAC/AEGL Committee evaluates the resultant AEGL values within the context of other supporting data to determine the reasonableness of the extrapolated values. This evaluation is done even when the value of n is derived from empirical data that describe the exposure concentration-duration relationship. The NAC/AEGL Committee uses a value for n that results in AEGL values that best fit the supporting data.

2.7.8.1 Use of Empirical Data to Determine the Exposure Concentration–Exposure Duration Relationship

The rationale for the selection of an empirically based time-scaling approach should include the following:

1. The health-effect endpoint of concern.
2. The exposure durations for which data were available.
3. Description of the statistical method used. If no method was used, then describe how the value of n was derived.
4. Description of the data used, including durations or the concentration-duration values used for extrapolation. Include the formula used.
5. Description of the different values of n from one or more studies and why a specific derived value of n was used.
6. The value of k calculated from $C^n \times t = k$ after uncertainty and/or modifying factors have been applied to C.
7. If the value of n is based on an analysis of the combined data from a

number of different studies, then provide a description of how the different concentration-duration values were combined and why they were used.

8. If definitive supporting data for 10-min exposures are not available when extrapolating from 8-h empirical data, the 10-min AEGL generally is assigned the same value as that extrapolated for the 30-min AEGL.

2.7.8.2 Estimating the Exposure Concentration–Exposure Duration Relationship Using a Structurally Related Material

The rationale for the selection of this time-scaling approach should include the following:

1. Description of the structure-activity relationships between the two chemicals.
2. The health-effect endpoint of concern.
3. The exposure durations for which data were available.
4. The statistical method used or a statement of how the value of n was derived.
5. Description of the data from the surrogate chemical used to derive the exposure concentration-duration relationship. If a derived value of n is used, the equation should be included.
6. A description of how the different concentration-duration values were combined and why they were used if the value of n is based on an analysis of the combined data from a number of studies.
7. The value of k calculated after uncertainty and modifying factors have been applied.

2.7.8.3 Estimating the Exposure Concentration–Exposure Duration Relationship When Data Are Not Available to Derive a Value for n and Supporting Data Are Available

In the absence of data to derive a value for n, a value for n of 1 is initially selected when extrapolating from shorter to longer exposure durations and a value for n of 3 is initially selected when extrapolating from longer to shorter exposure durations. The values of n are used with the equation $C^n \times t = k$ to extrapolate from the empirically reported exposure concentrations and exposure durations to the AEGL-specified exposure durations. AEGL values in accord with the supporting data are then selected.

The rationale for the selection of the time-scaling approach should include the following:

1. Presentation of the rationale in the technical support document (TSD) as follows: The relationship between concentration and duration of exposure as related to lethality was examined by ten Berge et al. (1986) for approximately 20 irritant or systemically acting vapors and gases. The authors subjected the individual animal data sets to probit analysis with exposure duration and exposure concentration as independent variables. An exponential function ($C^n \times t = k$), where the value of n ranged from 0.8 to 3.5 for different chemicals, was found to be an accurate quantitative descriptor for the chemicals evaluated. Approximately 90% of the values of n range between $n = 1$ and $n = 3$. Consequently, these values were selected as the reasonable lower and upper bounds of n. A value of $n = 1$ is used initially when extrapolating from shorter to longer periods, because the extrapolated values represent the most conservative approach in the absence of other data. Conversely, a value of $n = 3$ is used when extrapolating from longer to shorter periods, because the extrapolated values are more conservative in the absence of other data. If supporting data are available (description and references for data should be included) and indicate that the AEGL value initially extrapolated is inaccurate, the AEGL value has been adjusted to reflect the data.
2. Presentation of the AEGL values or exposure concentrations extrapolated from data using a value of $n = 1$ or $n = 3$ and the adjustments made as a result of supporting data..
3. Because of the uncertainty in 10-min exposure data and the absence of definitive supporting data, the 10-min AEGL is assigned the same value as the 30-min AEGL when extrapolating from 4-h or 8-h empirical data.
4. Discussion of the adjustments made and the rationale for making them.

2.7.8.4 Determining Exposure Concentration–Exposure Duration Relationships When Data Are Not Available to Derive a Value for n and No Supporting Data Are Available

In the absence of data to derive a value of n and the absence of supporting data to validate a value of n, the value of $n = 1$ will be selected for extrapolating from shorter to longer exposure durations, and the value n = 3 will be selected for extrapolating from longer to shorter exposure durations. If definitive supporting data for 10-min exposures are not available when extrapolating from 8-h empirical data, the 10-min AEGL generally is assigned the same value as that extrapolated for the 30-min AEGL.

The rationale for the selection of this time-scaling approach should include the following:

1. Presentation of the rationale in the TSD as follows: The relationship between concentration and duration of exposure as related to lethality was examined by ten Berge et al. (1986) for approximately 20 irritant or systemically acting vapors and gases. The authors subjected the individual animal data sets to probit analysis with exposure duration and exposure concentration as independent variables. An exponential function ($C^n \times t = k$), where the value of n ranged from 0.8 to 3.5 for different chemicals, was found to be an accurate quantitative descriptor for the chemicals evaluated. Approximately 90% of the values of n range between n = 1 and n = 3. Consequently, these values were selected as the reasonable lower and upper bounds of n to use when data are not available to derive a value of n. A value of n = 1 is used when extrapolating from shorter to longer periods, because the extrapolated values are conservative and therefore, reasonable in the absence of any data to the contrary. Conversely, a value of n = 3 is used when extrapolating from longer to shorter periods, because the extrapolated values are conservative and therefore reasonable in the absence of any data to the contrary. Because of the uncertainty in 10-min exposure data and the absence of definitive supporting data, the 10-min AEGL is assigned the same value as the 30-min AEGL when extrapolating from 4-h or 8-h empirical data.

2.7.8.5 AEGL Exposure Values Are Constant Across Time

The rationale for the selection of the time-scaling approach should include the following:

1. The data and mode or mechanism of action of the chemical and its effect on humans that supports the assignment of constant AEGL values across exposure durations.

2.8 GUIDELINES AND CRITERIA FOR ADDRESSING SHORT-TERM EXPOSURE TO KNOWN AND SUSPECT CARCINOGENS

Cancer represents a serious adverse health effect. Historically, the concerns for chemically induced cancers were based on continuous long-term exposure in controlled animal studies or information derived from clinical or epidemiologic studies of continuous or long-term exposures in humans. To conduct quantitative risk assessments for cancer in humans, mathematical (probit-log-dose) models were developed to utilize primarily animal bioassay data and extrapolate from the higher experimental levels to assess the carcinogenic risk to humans at low levels of chemical exposure. The evolution and usefulness of mathematical models to accommodate new understanding or new concepts regarding the mechanisms of carcinogenesis have been summarized in two publications by the NRC: *Guidelines for Developing Spacecraft Maximum Allowable Concentrations for Space Station Contaminants* (NRC 1992a), and *Guidelines for Developing Community Emergency Exposure Levels for Hazardous Substances* (NRC 1993a).

In the United States, some state and federal regulatory agencies conduct quantitative risk assessments on known or suspect carcinogens for continuous or long-term human exposure by extrapolating downward in linear fashion from an upper confidence limit on theoretical excess risk (FDA 1985; EPA 1986). The values derived for a specified "acceptable" theoretical excess risk to the U.S. human population, based on a lifetime of exposure to a carcinogenic substance, have been used extensively for regulatory purposes.

There are no adopted state or federal regulatory methodologies for deriving short-term exposure standards for workplace or ambient air based on carcinogenic risk, because nearly all carcinogenicity studies in animals and retrospective epidemiologic studies have entailed high-dose, long-term exposures. As a result, there is uncertainty regarding the extrapolation from continuous lifetime studies in animals to the case of once-in-a-lifetime human exposures. This is particularly problematical, because the specific biologic mechanisms at the molecular, cellular, and tissue levels leading to cancer are often exceedingly diverse, complex, or not known. It is also possible that the mechanisms of injury of brief, high-dose exposures will often differ from those following long-term exposures. To date, U.S. federal regulatory agencies have not established regulatory standards based on, or applicable to, less than lifetime exposures to carcinogenic substances.

2.8.1 NRC Guidance

Guidance on the development of short-term exposure levels, published by the NRC, identified cancer as one of the potential adverse health effects that might be associated with short-term inhalation exposures to certain chemical substances (NRC 1993a). That guidance document discusses and recommends specific risk-assessment methods for known genotoxic carcinogens and for carcinogens whose mechanisms are not well understood. As a first approximation, the default approach involves linear low-dose extrapolation from an upper confidence limit on theoretical excess risk. Further, the NRC guidance states that the determination of short-term exposure levels will require the translation of risks estimated from continuous long-term exposures to risks associated with short-term exposures. Conceptually, the approach recommended for genotoxic carcinogens adopted the method developed by Crump and Howe (1984) for applying the linearized multistage model to assessing carcinogenic risks based on exposures of short duration. In the case of nongenotoxic chemical carcinogens, the NRC guidance acknowledges that the approach is less clear because of the many different modes of action, the complexities of epigenetic carcinogenic mechanisms, and, in many cases, the paucity of data on chemical-specific mode of action. It is acknowledged also that dose thresholds exist for certain nongenotoxic carcinogens. The NRC guidance suggests that, in lieu of linear, low-dose extrapolation, approaches involving noncarcinogen risk-assessment techniques or cell kinetic models from the class of initiation-promotion-progression models be used, provided a known mechanism of action can justify the specific approach. The guidance emphasizes the importance of the underlying biologic processes when using any such models.

2.8.2 Precedents for Developing Short-Term Exposure Limits Based on Carcinogenicity

The NRC guidance (1993a) for assessing the excess risks of genotoxic carcinogens is based on an adaptation of the work of Crump and Howe (1984). The Committee on Toxicology's (COT's) adaptation of the method was made for developing emergency exposure guidance levels (EEGLs) and short-term public emergency guidance levels (SPEGLs) for the U.S. Department of Defense (NRC 1986). EEGLs represent exposure levels that are acceptable for the performance of specific tasks by military personnel during emergency conditions lasting 1 to 24 h. SPEGLs represent acceptable airborne concentrations for a single, unpredicted short-term emergency exposure to the public. The exposure periods range from 1 h or less to 24 h, and the SPEGLs are generally set at 0.1 to 0.5 times the corresponding EEGL values.

The criteria and methods document prepared by COT for the development of EEGLs and SPEGLs indicates that theoretical excess carcinogenic risk levels in the range of 10^{-4} to 10^{-6} were generally considered acceptable risk levels (NRC 1986). However, the document states, "The role of short-term exposures in producing cancer is not clear.... [A]ny exposure to a carcinogen has the potential to add to the probability of carcinogenic effects ... [but] the effects of long or repeated exposures could greatly overshadow brief exposures (up to 24 h)." Additionally, the NRC report states, "The assumption that the carcinogenic response is directly proportional to total dose is likely not to hold for all materials and all tissues that these materials affect." However, these concerns not withstanding, the NRC set SPEGL values based on the carcinogenic risk-assessment method previously mentioned for hydrazine, methyl hydrazine, and 1,1-dimethyl hydrazine (NRC 1985). In each case, the theoretical excess cancer risk level used was 10^{-4}, and the derived values were determined to be lower than corresponding airborne concentrations that were estimated to cause acute toxicity. SPEGL values for exposure periods of less than 24 h of other known or suspect human carcinogens were not based on carcinogenicity. These chemicals included benzene, trichloroethylene, ethylene oxide, and lithium chromate.

The National Aeronautics and Space Administration (NASA) requested that the NRC develop spacecraft maximum allowable concentrations (SMACs) for space-station contaminants. The NRC published guidelines for the development of short-term and long-term SMACs (NRC 1992a). Short-term SMACs refer to concentrations of airborne substances that will not compromise the performance of specific tasks during emergency conditions lasting up to 24 h. Because of NASA's concern for the health, safety, and functional abilities of space crews, SMACs for exposure from 1 to 24 h should not cause serious or permanent effects but may cause reversible effects that do not impair judgment or interfere with proper responses to emergencies. The long-term SMACs are designed to prevent deterioration in space-crew performance with continuous exposure for up to 180 days.

The guidelines for determining SMACs for carcinogens recommend the methods proposed by Kodell et. al. (1987) based on the linear multistage model. The level of theoretical excess risk used in the computation was 10^{-4}. The guidelines suggest extrapolations of long-term (often lifetime) exposures to shorter durations, such as 1, 30, or 180 days, and refer to a single-day exposure as "the case of near instantaneous exposure." Further, the guidance states, "It must be remembered that extrapolation from a daily lifetime exposure level and conversion to an instantaneous exposure level using ... [equations presented] ... is an extreme case and is valid only under the assumptions underlying the multistage theory of carcinogenesis." A review of the first three volumes of published SMACs (35 chemicals), including 10 known or suspected carcinogens, indicated that an assessment of excess risk for less than a 24-h exposure period was conducted on only 1 of the 10 carcinogenic substances. Carcinogenic assessments for excess risk were conducted on all 10 chemicals for 24 h, as well as 7, 30, and 180 days. The reasons provided in the NRC report for not undertaking a risk assessment on carcinogenic substances for exposure periods of less than 24 h included the following: (1) "Data not considered applicable to the exposure time (1 hr)"; (2) "Extrapolation to 1 hour exposure duration produces unacceptable uncertainty in the values"; and (3) "The NRC model was not used to calculate acceptable concentrations for exposures shorter than 24 hours" (NRC 1992a).

As stated previously, to date no U.S. federal or state regulatory agency has promulgated or established regulatory limits for single short-term (less than 24 h) exposures based on carcinogenic properties.

2.8.3 Scientific Basis for Credible Theoretical Excess Carcinogenic Risk Assessments for Single Exposures of 8 Hours or Less

The NRC guidance (NRC 1993a) explains that AEGLs can be developed using carcinogenic risk-assessment methods for exposure durations of 1 to 8 h provided adequate data are available. However, the guidance states that risk assessments on chemical carcinogenicity in humans should be based on all relevant data and embody sound biologic and statistical principles. While some of the substances may be considered known human carcinogens, most of the information is based on animal testing information. Additionally, since the mode of action for animal carcinogens are not always the same with respect to biologic properties among animal species or strains and humans, a weight-of-evidence evaluation must be carried out on a case-by-case basis. The weight-of-evidence evaluation considers comparative metabolic disposition, dose-dependent pharmacokinetic parameters, routes of exposure, mechanisms of action, and organ or species differences in response in animals and humans.

Uncertainties regarding lifetime theoretical excess carcinogenic risk assessments increase as shorter durations of a single exposure are considered. Most of these concerns stem from the reliance of both conclusions of carcinogenicity and quantitative assessments on long-term exposures in humans in occupational settings or in test animals. Thus, calculations for short-term risks require substantial extrapolation. At the same time, there are special concerns and unresolved issues regarding short exposures that will require more relevant data before they can be resolved. As evidenced from the actual application of these guidelines, COT was reluctant in most cases to develop quantitative carcinogenic risk assessments for less than 24-h exposures in the development of SMACs.

To better understand the empirical database for single exposures, EPA

DERIVATION OF AEGL VALUES 115

funded a study for the AEGL Program (Dr. Edward Calabrese of the University of Massachusetts was given a contract) to review the published literature and assess the circumstances during which a single exposure of short duration may be associated with a confirmed increase in carcinogenic response. This effort, referred to as the Single Exposure Carcinogen Database, represents a computerized summary that may assist the NAC/AEGL Committee's assessment of whether a single exposure to a particular chemical under consideration for AEGL development might cause tumor development following a one-time inhalation exposure. The data base is designed to contain numerous parameters important to tumor outcome and/or quality of the studies conducted. The database will contain approximately 5,500 "studies" or data sets involving approximately 500 chemicals from nearly 2,000 references.

Although a summary of the Single Exposure Carcinogen Database Project has been presented to the NAC/AEGL Committee, at the present time it is not known whether the data available on single exposures of carcinogenic substances will be sufficient to justify their use in the development of AEGL values. A preliminary review of the database indicates that only a limited number of short-term carcinogenesis bioassays conducted by the inhalation route are available. Hence, route-to-route extrapolations would be required in a manner that would not substantially weaken the conclusions reached for certain substances using standard EPA or NRC procedures if the toxicant is likely to cause tumors at a site other than the port of entry. If the substance causes tumors at the site of application or port of entry in oral or parenteral protocols, extrapolation to the inhalation route of exposure becomes problematic. For this reason, the NAC/AEGL Committee in most cases will rely on data from long-term animal studies as the basis for the quantitative cancer risk assessments for short-term exposures of 8 h or less. It is anticipated that work on the Single Carcinogen Database may be completed in 2001.

The Single Carcinogen Database may prove to be useful in obtaining some important information for AEGL development. The database shows that single exposure to various chemical classes, using various species and strains of animals, can result in tumor formation. Furthermore, chemicals can be selected from the database for which there is dose-response information. Data and information from positive responses of the chemical in the database could be compared between the single-dose study and the long-term study.

2.8.4 Practical Issues of Using Quantitative, Carcinogenic Risk Assessments for Developing AEGLs

In addition to fundamental scientific issues regarding carcinogenic risk assessments in the development of AEGL values, there are important practical issues to be considered by emergency planners and responders regarding AEGL values based on possible carcinogenic risk. The theoretical excess cancer risk for a lifetime exposure to known or suspect human carcinogens considered safe and protective of the public health ranges from 10^{-6} to 10^{-4} for EPA and most other U.S. federal regulatory agencies (EPA 1991). The AEGL values, however, are designed for emergency planning for, response to, and prevention of accidental releases from chemical accidents. Thus, theoretical excess cancer risk may be accumulated in 30 min or in a few hours. In addition to the individual risk of 10^{-6} to 10^{-4}, one should also consider a measure of population based risk. Experts in the chemical accident field indicate that the typical U.S. population at risk during most accidental chemical releases is in the range of 1,000 to 5,000 persons. The actual number of persons exposed depends on many factors, such as population density, quantity released, release rate, prevailing wind direction and velocity, terrain, and ambient temperature to name a few. Therefore, a population-based risk range of 10^{-6} to 10^{-4}, assuming a credible carcinogenic assessment can be made, approaches zero for a population of 1,000 to 5,000 or higher. The consideration of population-based risks using assessment methodologies designed for individual risks has precedent in EPA assessments of new industrial chemicals under TSCA (Toxic Substance Control Act) Section 5 and pesticide chemicals under FIFRA (Federal Insecticide Fungicide and Rodenticide Act).

Implementation of emergency-response procedures based on theoretical excess risk values of 10^{-6} to 10^{-4} may be problematical. For example, if such values were used, they would be based on an anticipated increased cancer risk of 10^{-6} to 10^{-4}, a policy consistent with EPA's acceptable cancer risk for lifetime exposures to known or suspect human carcinogens. However, the public health and safety risks associated with evacuation and other response measures might pose greater risks of injury or perhaps death. Thus, setting AEGL values based on uncertain theoretical cancer risk estimates might lead to response measures that increase actual or total risk for the exposed population.

2.8.5 Current Approach of the NAC/AEGL Committee to Assessing Potential Single Exposure Carcinogenic Risks

On the basis of the discussions and considerations presented in the earlier sections of this chapter on cancer risk assessment, the NAC/AEGL Committee has developed no AEGL values based on carcinogenicity. In view of the great uncertainty of the assumptions used in extrapolating from lifetime exposures to 8 h or less, the paucity of single-exposure inhalation data, the relatively small populations involved, and the potential risks associated with evacuations and other response measures, the NAC/AEGL Committee does not believe their use in setting AEGL values is justifiable at the present time.

However, the NAC/AEGL Committee will continue to identify and evaluate carcinogenic data during the development of AEGLs on a chemical-by-chemical basis. The scientific parameters used in this analysis are presented later in this section. In those cases in which, in the judgment of the committee, it is appropriate, risk assessments for 10^{-4}, 10^{-5}, and 10^{-6} levels of cancer risk will be conducted. It is believed that information on known or suspect human carcinogens should be provided to emergency planners and responders and made available to the public even when such information is not used to set AEGL values. Therefore, the NAC/AEGL Committee will continue to provide data and information on the carcinogenic properties of chemicals in the TSDs and, in instances in which the appropriate data are available, develop quantitative cancer risk assessments at risk levels of 10^{-4}, 10^{-5}, and 10^{-6} in accordance with the NRC guidance (NRC 1993a). The NAC/AEGL Committee will attempt to limit potential cancer risk to 10^{-4} or less when there are scientifically credible data to support the risk estimates when based on a single exposure. When substantial and convincing scientific data become available that clearly establish a relationship between a single short-term inhalation exposure to a chemical and the onset of tumors that are likely to occur in humans, the carcinogenic risk in the development of the appropriate AEGL values will be given appropriate weight-of-evidence considerations.

2.8.5.1 Evaluation of Carcinogenicity Data

The evaluation of the carcinogenic potential of a chemical exposure in humans must be based on analyses of all relevant data. Human epidemiologic and clinical studies, as well as accidental-exposure reports are considered and used to evaluate the carcinogenic potential of a substance. In the absence of human data, long-term bioassay data from controlled animal studies are used to derive theoretical excess carcinogenic risk estimates for exposed humans. The selection of data for estimating risk is based on the species and strain considered to resemble the human response most closely to provide the most accurate estimates.

Data suggestive of a single exposure inducing a carcinogenic response, including related mechanistic data that support such a possibility, are considered. Weight should be given to those studies most relevant to estimating effects in humans on a case-by-case basis. Data for assessing the strength of conclusions drawn from controlled animal studies should include information on comparative metabolic pathways, dose-dependent pharmacokinetic parameters, mode of exposure, mechanisms of action, and organ or species differences in response. In general, the NAC/AEGL Committee will follow a weight-of-evidence approach in the evaluation of carcinogenicity that is consistent with the availability and biologic variability of the data and its relationship to the likelihood of effects in humans.

2.8.5.2 Methodology Used for Assessing the Carcinogenic Risk of a Single Exposure

Guidance published by the NRC (1993a) states that the setting of AEGLs (CEELs) should involve linear low-dose extrapolation from an upper confidence limit on excess risk for genotoxic carcinogens and for carcinogens with mechanisms of action that are not well understood. More specifically, the NRC guidance suggests an approach utilizing the methods proposed by Kodell et al. (1987) based on multistage models. Although the NRC guidance states that multistage models could be useful for setting AEGL values, the guidance acknowledges that sufficient information may not be available to postulate the total number of stages in the cancer process and the stages that are dose-related. In these instances, the NRC guidance recommends the use of the time-weighted-average dose where the instantaneous dose D at time t_0 is assumed to be the equivalent of the lifetime excess carcinogenic risk as daily dose D up to time t. This equivalence is expressed by the equation $D = d \times t$. As shown by Kodell et al. (1987), the actual risk will not exceed the number of stages in the model (k). In instances in which multistage models can be used and prudence dictates conservatism, the NRC guidance suggests reducing the approximation of D by an adjustment factor of 2 to 6, depending on the number of assumed stages in the multistage model used.

To date, the NAC/AEGL Committee has evaluated excess theoretical risk at levels of 10^{-4}, 10^{-5}, and 10^{-6} for a one-time exposure to known or suspect human carcinogens by determining the total cumulative lifetime dose and applying Haber's law for exposure periods ranging from 30 min to 8 h. The resultant doses are then divided by an adjustment factor to account for the multistage nature of carcinogens (see Appendix H).

2.8.5.2.1 Determination of an Adjustment Factor for the Dose-Dependent Stage of Carcinogenesis

There is an extensive body of literature that deals with the concept of malignant tumor development, progression of an initiated cell through of successive stages, and quantitative carcinogenic risk assessment. Two references, Crump and Howe (1984) and Kodell et al. (1987), are cited in the NRC (1993a) publication. The concept has been further discussed (Goddard et al 1995; Murdoch et al. 1992; Murdoch and Krewski 1988; Bogen 1989). This process is referred to as a cell kinetic multistage model. There are several published variations of the basic tenants in the model. If only one or more stages are dose-dependent and exposure is concentrated in the dose-dependent stage, it is possible to underestimate risk when the risk is based on lifetime exposure. For example, if the first stage is dose-dependent, and there is a single exposure to an infant, the probability of cancer induction is maximized, because the entire lifetime of the individual is available for progression through the remaining stages in the development of the cancer. If the same dose were given to an elderly person, the probability of inducing cancer approaches zero, because there is insufficient time remaining in the life of that individual for the initiated cell to progress through the subsequent stages to a malignant cancer. Kodell et al. (1987) demonstrated that the underestimation of risk that is based on a lifetime of exposure will not exceed the number of stages in the multistage model. For this reason, the NRC (1986) recommends dividing the risk assessment based on the lifetime exposure by a factor between 2 and 6 to account for the number of stages in the multistage model applicable to the particular chemical of concern.

In addition to the multistage model, there have been a number of publications investigating the two-stage birth-death-mutation model (Morrison 1987; Chen et al. 1988; Murdoch and Krewski 1988; Moolgavkar and Luebeck 1990; Murdoch et al. 1992; Goddard et al. 1995). This model is similar to the multistage model. However, the impact of the number of stem cells at the time of chemical exposure is considered as well as the net growth rate of cells that have undergone the first stage of initiation. If the first-stage initiating event creates a cell that has a net growth rate greater than that of the stem cell, then the risk of that initiating event will be greater than it would be if the initiated cell grew at the same relative rate as the stem cell. In this case, exposure early in life results in a greater risk than exposure late in life. Conversely, exposure to a promoter (effects only the second stage) late in life will be more effective than early exposure, because relatively more initiated cells are present. If this stage is the only stage affected by the chemical, this situation is the same as that in the two stages of the multistage model. However, if the net growth rate of the initiated cells is 10 times the stem-cell rate, the relative effectiveness of exposure late in life could be 10-fold (Murdoch and Krewski 1988). Exposure to promoters between the first- and second-stage event can have an impact by increasing the net growth rate of initiated cells over that of stem cells. For maximum effectiveness, exposure to promoters (generally considered to be a nongenotoxic event) must encompass multiple challenge (Chen et al. 1988; Murdoch and Krewski 1988). Thus, the cancer risk associated with a single exposure to a promoter should not be greater than that predicted for multiple exposures, and no correction to the estimated risk has to be made in this case.

A major impact upon the risk assessment of the two-stage model comes from carcinogen exposure during the first stage in which the initiation event creates a cell with a greater net growth rate than the stem-cell rate. Modelers have considered a number of scenarios in which the net growth rate of initiated cells varies from -10 to +10. The greatest increase in risk in the two-stage model occurs when the first stage is dose-dependent and the initiating event creates a cell with a net growth rate of +10. In that case, the increased risk is 10-fold (Murdoch and Krewski 1988; Murdoch et al. 1992; Goddard et al. 1995).

Data are lacking on the biologic plausibility of the maximum value for the net growth rate of initiated cells (Murdoch et al 1992). Major data requirements for the two-stage birth-death-mutation model include the number of stem cells at different times of the life cycle, their rates of division and differentiation, and their response to chemical exposure in terms of cell division and mutation rate. This information is also needed for the initiated-cell populations (Moolgavkar and Luebeck 1990). Because of these data gaps, the projections made for the two-stage model remain more speculative than those for the multistage model in which there is general agreement that the number of stages should not exceed six.

For the above reasons, the linearized multistage model is used as a default when estimating risks for short-term exposures from lifetime carcinogenesis bioassays. In all of the above referenced publications on the multistage model, the maximum number of stages modeled was six.

AEGL values are applicable to humans in all stages of life, so the maximum risk to an infant must be taken into consideration. In this case, the concentration based on a lifetime exposure study is divided by 6 unless there is evidence that the chemical is a late-stage carcinogen or operates by mechanisms different from those assumed in development of the linearized multistage model. As a first approximation, the NAC/AEGL Committee will use the divisor of 6 in agreement with the NRC (1993a) guidance on the development of short-term exposure limits, which states that a factor of 6 represents a conservative adjustment factor for a near-instantaneous exposure.

2.8.5.3 Summary of Cancer-Assessment Methodology Used by the NAC/AEGL Committee

The EPA q_1^* values listed on the Integrated Risk Information System (IRIS) or the GLOBAL86-generated slope-factor values (Howe et al. 1986) are used to compute lifetime theoretical excess carcinogenic risk levels. These values are based on EPA (1986) guidance. The EPA (1996a) proposed methodology will be considered in the future. These values are used to compute the concentration for a single exposure for the time periods of interest. As discussed in the beginning of this section, these values are typically divided by 6 to account for early exposure to a carcinogen in which the first stage is dose-dependent or for late exposure to a carcinogen in which the last stage is dose-dependent. If there is information about the number of stages required for development of the cancer or the stage that is dose-dependent, the divisor will be modified accordingly. An example of a carcinogenicity assessment is given in Appendix H.

The cancer evaluation includes a weight-of-evidence discussion, which considers the following factors:

- Less evidence of carcinogenicity from a short-term exposure.
  — No evidence for human carcinogenicity (may or may not lend support to cancer induction from a single exposure but an important consideration).
  — Lifetime or long-term exposure necessary to elicit cancer.
  — Positive response only at very high doses.
  — Neoplasia appears reversible (when treatment is discontinued).
  — Appears to be a "threshold" carcinogen.
  — Weak or absent mutagenic response in multiple in vivo and in vitro test systems.
- Greater evidence of carcinogenicity from a short-term exposure.
  — Confirmed human carcinogen (may or may not lend support to cancer induction from a single exposure but an important consideration).
  — Short time to tumor.
  — Evidence for cancer from one to a few exposures.
  — Positive response at low doses.
  — Complete carcinogen.
  — Irreversible (when treatment is discontinued).
  — Potent mutagen in multiple in vivo and in vitro test systems.

2.9 GUIDELINES AND CRITERIA FOR MISCELLANEOUS PROCEDURES AND METHODS

2.9.1 Mathematical Rounding of AEGL Values

Given the uncertainties involved in generating AEGL values, it could be argued that only one significant figure should be used. However, because of a number of considerations discussed below, AEGL numbers are rounded to two significant figures (e.g., 1.5, 23, or 0.35). The value 7.35 is rounded to 7.4.

Trivial differences in numbers can give large differences in practice if only one significant figure is used. For example, values of 14.9 and 15.1 would yield AEGL values of 10 and 20, respectively. This is a 2-fold difference for a very small difference in computed AEGL values. Values of 18, 14, 11, and 6 ppm for 30 min, 1 h, 4 h, and 8 h would give values of 10, 10, 10, and 20 ppm, respectively, for the time points. As these numbers are often used in exposure models to make risk-management decisions, the use of two significant figures allows for a more reasonable progression when different exposure scenarios are considered.

Two significant figures may seem overly precise when values less than 1 ppm are presented, because those levels may be difficult to quantify to that level of precision. However, the AEGL-2 values will often be used to compare with ambient air-dispersion modeling projections for planning purposes. In this case, the use of two vs one significant figure can have substantial practical impact. Other rounding schemes may be used on a case-by-case basis with justification.

2.9.2 Multiplication of UFs

The NAC/AEGL Committee often multiplies two UFs of 3. Since the value 3 represents the geometric mean of 10 and 1, the actual number is 3.16. Therefore, the product of two different UFs is not 3 × 3 but 3.16 × 3.16, which equals 10. For simplicity's sake, 3.16 × 10 is represented by 30.

2.9.3 Conversion Between Parts per Million and Milligrams per Cubic Meter

Expressing the airborne concentration of a chemical in parts per million represents a volume-by-volume approach, and milligrams per cubic meter represents a mass-by-volume approach to quantifying the concentration. Because a change of temperature with little or no change in pressure or a change in pressure with little or no change in temperature will result in changes in the volume of the air with no change in the mass of the chemical dispersed in the air, the airborne concentration of the chemical expressed as milligrams per cubic meter can vary at different elevations above sea level and at different temperatures at the same elevation. Airborne concentrations expressed as parts per million represent a volume-by-volume comparison and therefore do not change, regardless of changes in elevation (pressure) or temperature.

AEGLs are expressed in parts per million. However, many inhalation studies on toxicity report the chemical concentrations in milligrams per cubic meter. In deriving AEGL values, it is assumed that the concentrations reported were measured at normal temperature and pressure (i.e., 25° C or 298° K and 760 mm Hg). The NAC/AEGL Committee uses this assumption in all cases in which concentrations in milligrams per cubic meter are converted from parts per million, or data in parts per million are converted to milligrams per cubic meter.

The effect of elevation above sea level is approximately 15% when comparing mass-by-volume (milligrams per cubic meter) in New York City and Denver, Colorado. Although elevation may not be a major consideration, the effect of converting parts per million to milligrams per cubic meter at various pressures (elevations) and temperatures can be made by using the following equation:

$$\text{mg/m}^3 \text{ at } P_a \text{ and } T_a = (\text{ppm}) \times ((MW)/(24.45 \times (760 \text{ mm Hg}) \times T_a/(P_a \times (298°K))))),$$

where

P_a = the absolute pressure (in mm Hg) at actual conditions.
T_a = the absolute temperature (in °K) at actual conditions.
MW = molecular weight.

3. Format and Content of Technical Support Documents

The technical support document (TSD) is the compilation of all relevant data and information from all key studies and references and the most important supporting studies and references for both human exposures and laboratory animal exposures. Additionally, this support document addresses all the methodologies used in the derivation of the AEGL values and presents the rationale and justification for the use of certain data in the derivation and for the elimination of certain studies or data. The TSD addresses why specific methodologies and adjustment factors were or were not used, the scientific evidence supporting the rationale and justification, and the appropriate references to the published scientific literature or sources of unpublished information.

Major components of the TSD are (1) a summary section that includes a concise summary of toxicity information on the chemical, rationales used for time scaling and selection of uncertainty factors, and a table of AEGL values for the three tiers as well as key references; (2) a detailed discussion of the items listed in 1; and (3) a derivation summary table that includes a list and discussion of the key data elements and the rationale used to derive the AEGL values.

EDITORIAL CONVENTIONS

- Concentrations will be expressed in the units used in the original publication. If the data in the publication or other data sources are

expressed in parts per million (ppm), enter only ppm values. If the data are expressed in milligrams per cubic meter (mg/m^3) or other units, then state the concentrations as expressed in the data source and add the ppm values in parentheses.
- References to footnotes should be superscript and lower case.

3.1 FORMAT AND CONTENT OF TECHNICAL SUPPORT DOCUMENTS

Preface

The AEGL tiers are defined in the Preface of each TSD. See Section 2.1 for definitions of AEGLs 1, 2, and 3.

Table of Contents

Major headings in the text, tables, and figures should be marked with the word processor indexing tool so that the Table of Contents can be generated by the computer. A sample Table of Contents is presented in Appendix D.

Summary

The Summary should include the following:

- The name and CAS number of the chemical being reviewed.
- A brief description of the substance, its physical properties, and uses.
- A brief statement or overview of the toxicology, including the extent of the data and information retrieved and reviewed, highlights of the most important research and strengths and weaknesses of the database. Discuss data on human exposures and data on laboratory animals.
- A brief summary (one paragraph for each AEGL tier) of the key study (with references), the data used, and the derivation of the AEGL values. Each summary will include the following:

  — Information on the endpoints of concern and exposure levels used as the basis for deriving the AEGL values.

- Exposure level (If the data in the publication are expressed in ppm, enter only ppm values. If the data are expressed in mg/m^3 or other units, state the concentrations as expressed in the publication and add the ppm values in parentheses.).
  - Exposure period.
  - Why the time-concentration point was selected (include effects observed or not observed, relate to the AEGL level, etc.).
  - The species and number of animals used.
  - Consistency with human data as appropriate.
  - The reference to the key study.
  - Uncertainty factors and modifying factors used or not used and why a specific value was chosen.
  - The time-scaling method used and why it was selected (include the rationale for the value of n in the time-scaling equation).
- A brief statement regarding carcinogenicity, if appropriate.
- A brief statement on the adequacy of the data (see Section 2.3.3 of this SOP manual).
- A summary table of draft/proposed AEGL values with
  - Values presented in ppm with mg/m^3 values in parentheses.
  - A rationale and reference for AEGL-1, -2, and -3.
  - Reasons for no AEGL value.
- References.

A sample Summary is presented in Appendix E.

Outline of the Main Body of the Technical Support Document

1. Introduction

- General information regarding occurrence, production and use, and physical-chemical data (table for physical-chemical data)

2. Human Toxicity Data

2.1 Acute Lethality (include anecdotal case reports if pertinent)
2.2 Nonlethal Toxicity
    2.2.1   Acute Studies (include anecdotal case reports if pertinent)
    2.2.2   Epidemiologic Studies
2.3 Developmental and Reproductive Toxicity

2.4 Genotoxicity
2.5 Carcinogenicity (include EPA and IARC classifications)
2.6 Summary (weight-of-evidence approach)

- Tabulation of data as appropriate within sections and/or in summary

3. Animal Toxicity Data

3.1 Acute Lethality (include species and strain, number of animals, exposure concentrations and durations, mortality rates and ratios, time to death) (To maintain a standardized format, the order of animals shown below should be used.)
  3.1.1   Nonhuman Primates
  3.1.2   Dogs
  3.1.3   Rats
  3.1.4   Mice
  3.1.5   Guinea Pigs
  3.1.6   Rabbits
  3.1.7   Other Species

- Sections to include relevant studies (potential key studies and supporting data) or provide overall picture of toxicity data as appropriate
- Third-level headers to vary, depending upon available data; exclusion of header to mean no relevant data available

3.2 Nonlethal Toxicity (include species and strain, number of animals, exposure concentrations and durations, critical effects, time-course data) (The order of animals shown should be used. If no data are available for a species, the number should be used for the next species discussed.)
  3.2.1   Nonhuman Primates
  3.2.2   Dogs
  3.2.3   Rats
  3.2.4   Mice
  3.2.5   Guinea Pigs
  3.2.6   Rabbits
  3.2.7   Other Species

- Sections to include relevant studies (potential key studies and supporting data) or provide overall picture of toxicity data as appropriate

- Third-level headers to vary, depending upon available data; exclusion of header to mean no relevant data available

3.3 Developmental and Reproductive Toxicity
3.4 Genotoxicity
3.5 Carcinogenicity
3.6 Summary (weight-of-evidence approach)

- Tabulation of data as appropriate within sections and/or in summary

4. Special Considerations

4.1 Metabolism and Disposition (general background; interspecies and individual variabilities, especially as they pertain to AEGL derivation)
4.2 Mechanism of Toxicity (general background; interspecies and individual variabilities, especially as they pertain to AEGL derivation)
4.3 Structure-Activity Relationships (data relevant to filling data gaps on the chemical)
4.4 Other Relevant Information
    4.4.1 Species Variability
    4.4.2 Concurrent Exposure Issues (e.g., potentiation)

- Third-level headers to vary, depending upon available data; exclusion of header to mean no relevant data available

5. Data Analysis for Proposed AEGL-1

5.1 Summary of Human Data Relevant to AEGL-1 (general summary description of selected key and supporting study or studies if available)
5.2 Summary of Animal Data Relevant to AEGL-1 (general summary description of selected key and supporting study or studies if available)
5.3 Derivation of AEGL-1 (key study, critical effect, dose-exposure concentration, uncertainty factor application and justification, temporal extrapolation, assumptions, confidence, consistency with human data if appropriate)

6. Data Analysis for Proposed AEGL-2

6.1 Summary of Human Data Relevant to AEGL-2 (general summary description of selected key and supporting study or studies if available)

6.2 Summary of Animal Data Relevant to AEGL-2 (general summary description of selected key and supporting study or studies if available)

6.3 Derivation of AEGL-2 (key study, critical effect, dose-exposure concentration, uncertainty factor application and justification, temporal extrapolation, assumptions, confidence, consistency with human data if appropriate)

7. Data Analysis for Proposed AEGL-3

7.1 Summary of Human Data Relevant to AEGL-3 (general summary description of selected key and supporting study or studies if available)

7.2 Summary of Animal Data Relevant to AEGL-3 (general summary description of selected key and supporting study or studies if available)

7.3 Derivation of AEGL-3 (key study, critical effect, dose-exposure concentration, uncertainty factor application and justification, temporal extrapolation, assumptions, confidence, consistency with human data if appropriate)

8. Summary of Proposed AEGLs

8.1 AEGL Values and Toxicity Endpoints

8.2 Comparison with Other Standards and Criteria (summarized in text and presented in a table; see SOP Appendix J for an example)

8.3 Data Adequacy and Research Needs (for content, see Section 2.3.3 of this manual)

9. References

- List of references cited in document.

10. Appendixes

APPENDIX A   (Derivation of AEGL Values) (see SOP Appendix F for an example)

APPENDIX B   (Time-Scaling Calculations) (see SOP Appendix G for an example)

APPENDIX C   (Carcinogenicity Assessment) (see SOP Appendix H for an example)

APPENDIX D   (Derivation Summary) (see SOP Appendix I for specific format and an example)
Format for Derivation Summary:

DERIVATION SUMMARY (CHEMICAL NAME)
(CAS No.)

AEGL-1 (OR -2 OR -3) Values				
10 min	30 min	1 h	4 h	8 h
ppm	ppm	ppm	ppm	ppm
Reference:				
Test Species/Strain/Number:				
Exposure Route/Concentrations/Durations:				
Effects:				
Endpoint/Concentration/Rationale:				
Uncertainty Factors/Rationale:				
Modifying Factor:				
Animal to Human Dosimetric Adjustment:				
Time Scaling:				
Data Adequacy:[a]				

[a] Elements that should be included in the Data Adequacy entry are discussed in Section 2.3.3 of this SOP manual. If an AEGL-1 value is not recommended, there should be a short discussion of the rationale for that choice. The rationale should include, as appropriate, a discussion that numeric values for AEGL-1 are not recommended because (1) relevant data are lacking, (2) the margin of safety between the derived AEGL-1 and AEGL-2 values is inadequate, or (3) the derived AEGL-1 is greater than the AEGL-2. Absence of an AEGL-1 does not imply that exposure below the AEGL-2 is without adverse effects.

3.2 GRAPHIC DESCRIPTION OF DATA

Graphic descriptions of relevant data can be helpful in identifying, understanding, and comparing similarities and differences, degree of variation, and trends among the values cited. Well-prepared graphs provide the reader with an overview of dose-response relationships in terms of both airborne concentrations and exposure periods in various studies and various species. The graphs should supplement the data tables but cannot replace tabular summaries. The graphs can be placed in the body of the document or in an appendix. Below are examples of presentations of graphic data.

Comparisons between different times and the toxicity values are difficult because the values vary according to the time. A particularly useful way to present the data (Table 3-1 and Figure 3-1) is based on the concept of placing the toxic response into severity categories (Hertzberg and Miller, 1985; Hertzberg and Wymer, 1991; Guth et al., 1991). In Table 3-1, the severity categories fit into definitions of the AEGL health effects. The category severity definitions for the column headings are 0 = no effect; 1 = discomfort; 2 = disabling; 3 = lethal; NL = did not die at a lethal concentration (at an experimental concentration in which some of the animals died and some did not, the NL label refers to the animals that did not die); AEGL or C = AEGL or censored (severity category could not be established). The effects that place an experimental result into a particular category vary according to the spectrum of data available on a specific chemical and the effects from exposure to that chemical. When the exposure concentration is placed into the appropriate column, the graph in Figure 3-1 is generated. The doses often span a number of orders of magnitude, especially when human data exist. Therefore, the concentration is placed on a log scale. Note that the AEGL values are designated as a triangle without an indication to their level. The AEGL-3 is higher than the AEGL-2, which is higher than the AEGL-1.

This type of plot is useful for a number of reasons and can be used to address the following questions:

- Are the AEGL values protective?
  — Are the AEGL-3 values below the concentration causing death in experimental animals? If the answer is no, then the question should be raised about the appropriateness of the AEGL-3 value. Is the AEGL-3 value appropriate and the data point anomalous, or should the AEGL-3 value be lowered?
  — Similar questions should be asked about the AEGL-1 and AEGL-2 values.

TABLE 3-1 Grouping Data into Categories for Plotting

Reference	Species	Sex	# Exposures	ppm	Min	ppm 0	ppm 1	ppm 2	ppm NL	ppm 3	ppm	Category	Gp Size	Incidence	Comments
Toluene											AEGL				Category
NAC/AEGL-1				115	30						115	AEGL			0 = no effect
NAC/AEGL-1				82	60						82	AEGL			1 = discomfort
NAC/AEGL-1				41	240						41	AEGL			2 = disabling
NAC/AEGL-1				29	480						29	AEGL			3 = lethal
NAC/AEGL-2				267	30						267	AEGL			ND = did not die at lethal concentration
NAC/AEGL-2				189	60						189	AEGL			AEGL or C = AEGL or censored
NAC/AEGL-2				94	240						94	AEGL			
NAC/AEGL-2				67	480						67	AEGL			
NAC/AEGL-3				897	30						897	AEGL			
NAC/AEGL-3				634	60						634	AEGL			
NAC/AEGL-3				317	240						317	AEGL			
NAC/AEGL-3				224	480						224	AEGL			
Baelum et al. 1985	hu		1	100	390		100					1			sensory irritation, sleepiness, intoxication, decreased manual dexterity and color discrimination
Wilson 1943	hu		many	200	480		200					1			headache, lassitude, anorexia
Wilson 1943	hu		many	200	480			200				2			headache, nausea, incoordination, increased reaction time

Reference	species	N							Effects
Wilson 1943	hu	many	500	480		500		2	headache, nausea, incoordination, increased reaction time
Wilson 1943	hu	many	500	480	500			1	headache, nausea, incoordination, decreased reaction time and palpitation, extreme weakness
Ukai et al. 1993	hu	many	100	480	100			1	weight loss, dizziness, headache, tightness in chest, dimmed vision
Lee et al. 1988	hu	many	100	480	100			1	weight loss, dizziness, headache
Gamberale & Hultengren 1972	hu	1	300	20	300			1	reaction time affected
Gamberale & Hultengren 1972	hu	1	700	20	700			1	decrease in perceptual speed
von Oettingen et al. 1942	hu	1	200	480		200		2	muscular weakness, confusion, impaired coordination, dilated pupils
von Oettingen et al. 1942	hu	1	200	480		200		2	severe incoordination, confusion, dilated pupils, nausea, extreme fatigue
von Oettingen et al. 1942	hu	1	600	480		600		2	severe incoordination, confusion, dilated pupils, nausea, extreme fatigue
von Oettingen et al. 1942	hu	1	800	180		800		2	loss of self-control, muscular weakness, extreme fatigue, nausea, bone marrow suppression

134

TABLE 3-1 (Continued)

Reference	Species	Sex	# Exposures	ppm	Min	ppm 0	ppm 1	ppm 2	ppm NL	ppm 3	ppm AEGL	Category	Gp Size	Incidence	Comments Category
Toluene															
Baelum et al. 1990	hu		1	100	420		100					1			sensory irritation, altered temp. perception, headache, dizziness
Echeverria et al. 1991	hu		1	150	420		150					1			decreased performance on spatial and neurobehavioral tasks, headache, eye irritation
Andersen et al. 1983	hu		1	40	360		40					1			no effect/sensory irritation, increase in odor level
Andersen et al. 1983	hu		1	100	360		100					1			no effect/sensory irritation, increase in odor level
Rahill et al. 1996	hu		1	100	360	100						0			increased latency on a neurobehavioral task (not a biologically relevant neurobehavioral deficit)
Dick et al. 1984	hu		1	100	240	100						0			decreased accuracy on visual-vigilance test (not a biologically relevant neurobehavioral deficit)
Cherry et al. 1983	hu		1	80	240	80						0			no impairment on neurobehavioral tasks
Carpenter et al. 1976	hu		1	220	15		220					1			sensory threshold for eye irritation

Reference	Species									
Pryor et al. 1978	rat	1	26,700	60			26,700		NL	LC50
Pryor et al. 1978	rat	1	26,700	60		26,700			3	LC50
Cameron et al. 1938	rat	1	24,400	90		24,400			NL	60% mortality
Cameron et al. 1938	rat	1	24,400	90			24,400		3	60% mortality
Kojima and Kobayashi 1973	rat	1	15,000	150		15,000			NL	80% mortality
Kojima and Kobayashi 1973	rat	1	15,000	150			15,000		3	80% mortality
Cameron et al. 1938	rat	1	12,200	390			12,200		3	100% mortality
Carpenter et al. 1976	rat	1	8,800	240		8,800			NL	LC50
Carpenter et al. 1976	rat	1	8,800	240			8,800		3	LC50
Smyth et al. 1969	rat	1	4,000	240		4,000			NL	16% mortality
Smyth et al. 1969	rat	1	4,000	240			4,000		3	16% mortality
Bonnet et al. 1979	mouse	1	6,940	360		6,940			NL	LC50
Bonnet et al. 1979	mouse	1	6,940	360			6,940		3	LC50
Svirbely et al. 1943	mouse	1	5,320	420		5,320			NL	LC50

TABLE 3-1 *(Continued)*

Toluene

Reference	Species	Sex	# Exposures	ppm	Min	ppm 0	ppm 1	ppm 2	ppm NL	ppm 3	ppm AEGL	Category	Gp Size	Incidence	Comments Category
Svirbely et al. 1943	mouse		1	5,320	420					5,320		3			LC50
Moser and Balster 1985	mouse		1	38,465	10				38,465			NL			LC50
Moser and Balster 1985	mouse		1	38,465	10					38,465		3			LC50
Moser and Balster 1985	mouse		1	21,872	30				21,872			NL			LC50
Moser and Balster 1985	mouse		1	21,872	30					21,872		3			LC50
Moser and Balster 1985	mouse		1	19,018	60				19018			NL			LC50
Moser and Balster 1985	mouse		1	19,018	60					19,018		3			LC50

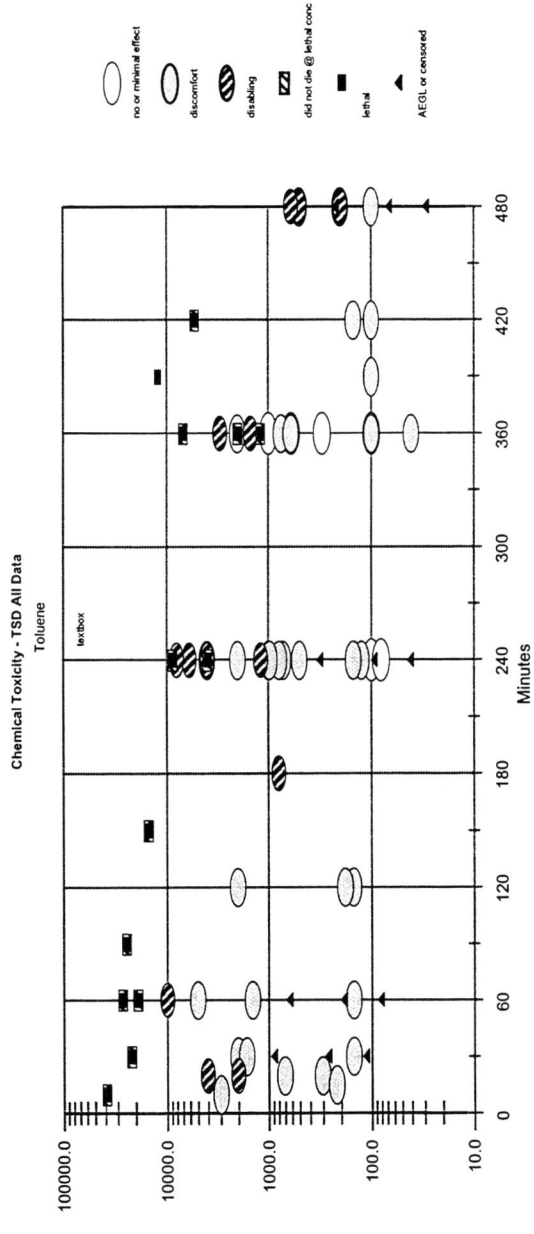

FIGURE 3-1 Plot of categories of data.

- Are there data points that appear to be statistical outliers? Why are they outliers? Should they be considered in the development of AEGL values or discarded because of faulty experimental technique?
- Does the spread of data points for a particular severity category indicate major differences between species or are the results from different species congruent?
- Is the time-scaling algorithm reasonable and consistent with the data? For example, using the derived or chosen value of n in the equation $C^n \times t = k$, does the plot of the AEGL-3 values parallel the slope of the lethality data? Similar questions can be asked about the AEGL-1 and AEGL-2 plots.
- Is there evidence that a different time-scaling factor should be used for the AEGL-2?
- What are the most appropriate data points to use for the time scaling?

4. Current Administrative Processes and Procedures for the Development of AEGL Values

The primary purpose of the AEGL program and the NAC/AEGL Committee is to develop AEGLs for short-term exposures to airborne concentrations of acutely toxic, high-priority chemicals. AEGLs are needed for a wide range of planning, response, and prevention applications. These applications may include many U.S. initiatives, such as the Environmental Protection Agency (EPA) Superfund Amendments and Reauthorization Act (SARA) Title III Section 302-304 emergency planning program, the Clean Air Act Amendments (CAAA) Section 112(r) accident prevention program, and the remediation of Superfund sites program; the Department of Energy (DOE) environmental restoration, waste management, waste transport, and fixed facility programs; the Department of Transportation (DOT) emergency waste response program; the Department of Defense (DOD) environmental restoration, waste management, and fixed facility programs; the Agency for Toxic Substances and Disease Registry (ATSDR) health consultation and risk assessment programs; the National Institute for Occupational Safety and Health (NIOSH) and Occupational Safety and Health Administration (OSHA) regulations and guidelines for workplace exposure; the state CAA Section 112(b) programs and other state programs; the U.S. Chemical Manufacturer's Association (now known as American Chemistry Council) Chemtrec program; and other chemical emergency programs in the U.S. private sector. From an international perspective, it is anticipated that the AEGLs will find a wide range of applications in chemical emergency planning, response, and prevention programs in both the public and private sectors of member countries of the Organization for Economic Cooperation and Development (OECD). It is hoped that the AEGLs also will be used by other countries in the international community.

A principal objective of the NAC/AEGL Committee is to develop the most scientifically credible, acute (short-term) exposure guideline levels possible within the constraints of data availability, resources, and time. This objective includes highly effective and efficient efforts in data gathering, data evaluation, and data summarization, fostering the participation of a large cross-section of the relevant scientific community, both nationally and internationally, and the adoption of procedures and methodologies that facilitate consensus-building for AEGL values within the NAC/AEGL Committee.

Another principal objective of the NAC/AEGL Committee is to develop AEGL values for approximately 400 to 500 acutely hazardous substances within the next 10 years. Therefore, the near-term objective is to increase the production of AEGL documents to approximately 50 chemicals per year without exceeding budgetary limitations or compromising the scientific credibility of the values developed.

To reach these objectives, the NAC/AEGL Committee must adopt and adhere to specific processes and procedures both scientifically and administratively. This objective is accomplished through the development and maintenance of a comprehensive standing operating procedures (SOP) manual that addresses the scientific and administrative procedures required to achieve the objectives of the NAC/AEGL Committee. This section is devoted to those administrative processes and procedures deemed necessary to achieve the AEGL program objectives.

4.1 COMMITTEE MEMBERSHIP AND ORGANIZATIONAL STRUCTURE

The NAC/AEGL Committee is composed of representatives of U.S. federal, state and local agencies, and organizations in the private sector that derive programmatic or operational benefits from of the AEGL values. Federal representatives are from EPA, DOE, ATSDR, NIOSH, OSHA, DOT, DOD, the Centers for Disease Control and Prevention (CDC), the Food and Drug Administration (FDA), and the Federal Emergency Management Agency (FEMA). States providing committee representatives include New York, New Jersey, Texas, California, Minnesota, Illinois, Connecticut, and Vermont. Private companies with representatives include Honeywell, Inc., ExxonMobil, and Arch Chemical, Inc. Other organizations with representatives include the

American Industrial Hygiene Association (AIHA), the American College of Occupational and Environmental Medicine (ACOEM), the American Association of Poison Control Centers (AAPCC), and the American Federation of Labor–Congress of Industrial Organizations (AFL-CIO). In addition, the committee membership includes individuals from academia, a representative of environmental justice groups, and representatives of other organizations in the private sector. A current list of the NAC/AEGL Committee members and their affiliations is shown on page 5. At present, the committee has 31 members.

Recently, the Organization for Economic Cooperation and Development (OECD) and various OECD member countries have expressed an interest in the AEGL program. Several OECD member countries such as Germany and the Netherlands have been participating in committee activities and actively pursuing formal membership on the NAC/AEGL Committee. It is envisioned that the committee and the AEGL program in general will progressively expand its scope and participation to include the international community.

The director of the AEGL program has the overall responsibility for the entire AEGL program and the NAC/AEGL Committee and its activities. A designated federal officer (DFO) is responsible for all administrative matters related to the committee to ensure that it functions properly and efficiently. These individuals are not voting members of the committee. The NAC/AEGL Committee chair is appointed by EPA and is selected from among the committee members. In concert with the program director and the DFO, the chair coordinates the activities of the committee and also directs all formal meetings of the committee. From time to time, the members of the committee serve as chemical managers and chemical reviewers in a collaborative effort with assigned scientist authors (noncommittee members) to develop AEGLs for a specific chemical. These groups of individuals are referred to as AEGL Development Teams, and their function is discussed in Section 4.8 of this manual.

4.2 THE AEGL DEVELOPMENT AND PEER-REVIEW PROCESS

The process that has been established for the development of the AEGL values is the most comprehensive ever used for the determination of short-term exposure limits for acutely toxic chemicals. A summary of the overall process is presented in diagram form in Figure 4-1. The process consists of four basic stages in the development and status of the AEGLs, and they are identified according to the review level and concurrent status of the AEGL values. They include (1) "draft" AEGLs, (2) "proposed" AEGLs, (3) "interim" AEGLs, and (4) "final" AEGLs. The entire development process can

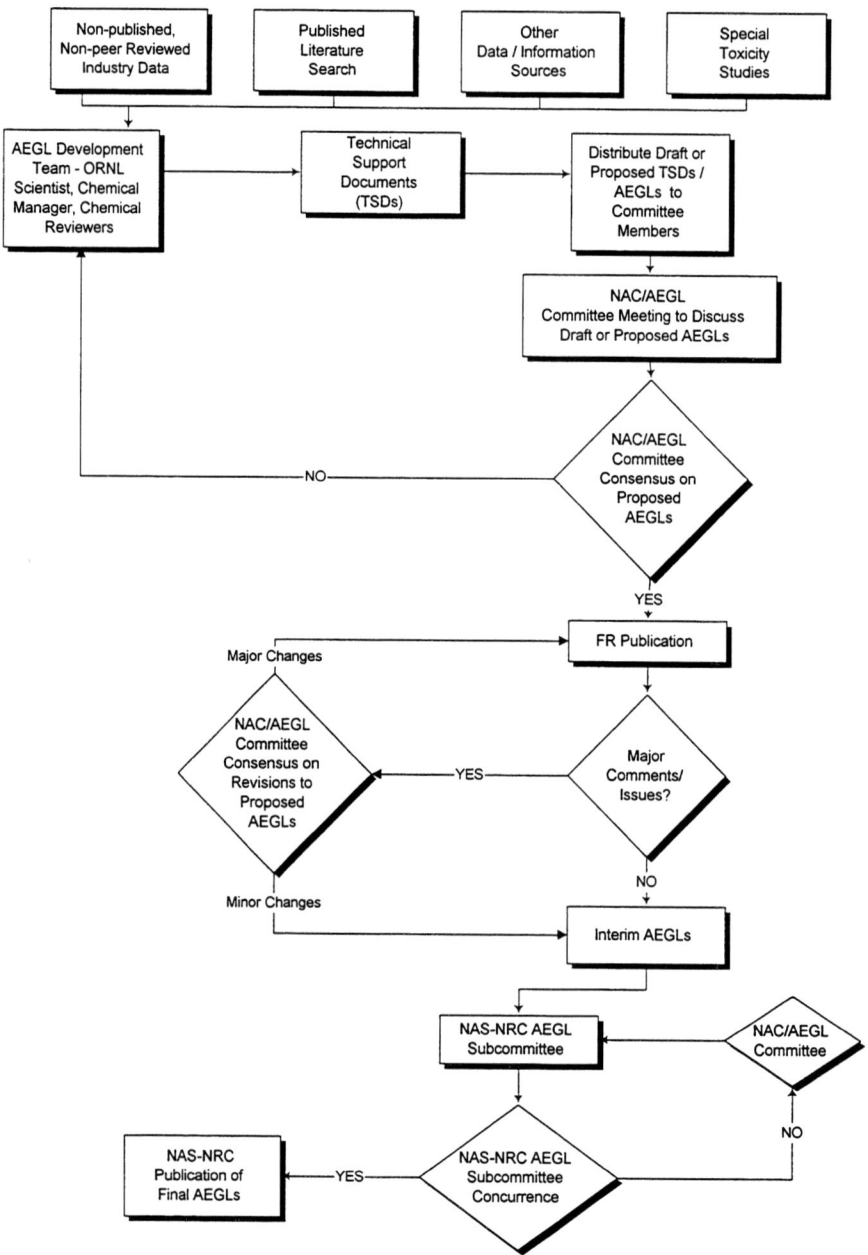

FIGURE 4-1 The AEGL development process.

be described by individually describing the four basic stages in the development of AEGL values.

Stage 1: Draft AEGLs

This first stage begins with a comprehensive search of the published scientific literature. Attempts are made to mobilize all relevant unpublished data through industry trade associations and from individual companies in the private sector. A more detailed description of the published and unpublished sources of data and information utilized is provided in Section 2.3 of this manual, which addresses search strategies. The data are evaluated following the guidelines published in the National Research Council (NRC) guidance document (NRC 1993a) and this SOP manual, and selected data are used as the basis for the derivation of the AEGL values and the supporting scientific rationale. Data evaluation, data selection, and the development of a TSD are all performed as a collaborative effort among the staff scientists at the organization that drafts TSDs, the chemical manager, and two chemical reviewers. This group is the AEGL Development Team. NAC/AEGL Committee members are specifically assigned this responsibility for each chemical under review. Hence, a separate team composed of different committee members is formed for each chemical under review. The product of this effort is a TSD that contains draft AEGLs. The draft TSD is subsequently circulated to all other NAC/AEGL Committee members for review and comment prior to a formal meeting of the committee. Revisions to the initial TSD and the draft AEGLs are made up to the time of the NAC/AEGL Committee meeting scheduled for formal presentation and discussion of the AEGL values and the documents. At the committee meeting, the committee deliberates and, if a quorum is present, attempts to reach a consensus or a two-thirds majority vote to elevate the AEGLs to "proposed" status. A quorum of the NAC/AEGL Committee is defined as 51% or more of the total NAC/AEGL Committee membership. If agreement cannot be reached, the committee conveys its issues and concerns to the AEGL Development Team, and further work is conducted by this group. After completion of additional work, the chemical is resubmitted for consideration at a future meeting. If a consensus or two-thirds majority vote of the committee cannot be achieved because of inadequate data, no AEGL values will be developed until adequate data become available.

Stage 2: Proposed AEGLs

Once the NAC/AEGL Committee has reached a consensus or a two-thirds majority vote on the AEGL values and supporting rationale, they are referred to as proposed AEGLs and are published in the *Federal Register* for a 30-day review and comment period. Following publication of the proposed AEGLs in the *Federal Register*, the committee reviews the public comments, addresses and resolves relevant issues, and seeks a consensus or a two-thirds majority vote on the original or modified AEGL values and the accompanying scientific rationale.

Stage 3: Interim AEGLs

Following resolution of relevant issues raised through public review and comment and subsequent approval of the committee, the AEGL values are classified as interim. The interim AEGL status represents the best efforts of the NAC/AEGL Committee to establish exposure limits, and the values are available for use as deemed appropriate on an interim basis by federal and state regulatory agencies and the private sector. The interim AEGLs, the supporting scientific rationale, and the TSD are subsequently presented to the NRC Subcommittee on Acute Exposure Guideline Levels (NRC/AEGL Subcommittee) for its review and concurrence. If concurrence cannot be achieved, the NRC/AEGL Subcommittee will submit its issues and concerns to the NAC/AEGL Committee for further work and resolution.

Stage 4: Final AEGLs

When concurrence by the NRC/AEGL Subcommittee is achieved, the AEGL values are considered final and published by the NRC. Final AEGLs may be used on a permanent basis by all federal, state, and local agencies and private-sector organizations. It is possible that from time to time new data will become available that challenges the scientific credibility of final AEGLs. If that occurs, the chemical will be resubmitted to the NRC/AEGL Subcommittee and recycled through the review process.

4.3 OPERATION OF THE NAC/AEGL COMMITTEE

The NAC/AEGL Committee meets formally four times each year for 2½ days. The meetings are scheduled for each quarter of the calendar year and are generally held in the months of March, June, September, and December. Based on overall cost considerations, the meetings are generally held in Washington, D.C. However, from time to time, committee meetings may be held at other locations for justifiable reasons.

At least 15 days prior to the committee meetings, a notice of the meeting is published in the *Federal Register* together with a list of chemicals and other matters to be addressed by the committee and provides dates, times, and location of the meetings. The agenda is finalized and distributed to committee members approximately 1 week prior to the meeting. The agenda also is available to other interested parties at that time, upon request, through the designated federal officer (DFO).

All NAC/AEGL Committee meetings are open to the public and interested parties may schedule individual presentations of relevant data and information by contacting the DFO to establish a date and time. Relevant data and information from interested parties also may be provided to the committee through the DFO during the period of development of the draft AEGLs so that it can be considered during the early stage of development. Data and information may be submitted during the proposed and interim stages of AEGL development as well.

The NAC/AEGL Committee meetings are conducted by the chair who is appointed by EPA in accordance with the Federal Advisory Committee Act (FACA). At the time of the meeting, both the chair and all other committee members will have received the initial draft and one or more revisions of the TSD and draft, proposed, or interim AEGL values for each chemical on the agenda. Reviews, comments, and revisions are continuous up to the time of the meeting, and committee members are expected to be familiar with the draft, proposed, or interim AEGLs, supporting rationale, and other data and information in each TSD and to participate in the resolution of residual issues at the meeting. Procedures for the AEGL Development Teams and the other committee members regarding work on AEGLs in proposed or interim status are similar to those for draft AEGLs.

All decisions of the NAC/AEGL Committee related to the development of draft, proposed, interim, and final AEGLs and their supporting rationale are made by consensus or a two-thirds majority vote of members at a committee meeting.

The highlights of each meeting are recorded by the scientists who draft the TSDs, and written minutes are prepared, ratified, and maintained in the committee's permanent records. Deliberations of each meeting also are tape-recorded and stored in the committee's permanent records by the DFO for future reference as necessary.

All proposed AEGL values and supporting scientific rationale are published in the *Federal Register*. Review and comment by interested parties and the general public are requested and encouraged. The committee's response to official comments on *Federal Register* Notices on proposed AEGL values consists of the discussions and deliberations that take place during the committee meetings for elevating the AEGLs from proposed to interim status. This information is reflected on the tapes and in the minutes of the meetings and will be maintained for future reference. Changes in the proposed AEGL values and the supporting rationale that are considered appropriate by the NAC/AEGL Committee based on *Federal Register* comments will be made prior to elevating the AEGLs to interim status.

As mentioned in Chapter 1, a Standing Operating Procedures Workgroup (SOP Workgroup), established in March 1997, documents, summarizes, and evaluates the various procedures, methodologies, and guidelines used by the committee in the gathering and evaluation of scientific data and information and in the development of the AEGL values. The SOP Workgroup performs a critical function by continually providing the committee with detailed information on the committee's interpretation of the NRC guidelines and on its approaches used in the derivation of each AEGL value for each chemical addressed. This documentation enables the committee to assess continually the basis for its decision-making, ensure consistency with the NRC guidelines, and maintain the scientific credibility of the AEGL values and accompanying scientific rationale. This ongoing effort is continually documented in the SOP manual.

4.4 ROLE OF THE DIRECTOR OF THE AEGL PROGRAM

The director has the overall responsibility for the AEGL program, including the NAC/AEGL Committee and its interface with other programs and organizations in the public- and private-sectors nationally and internationally. More specifically, he or she is responsible for the overall management of the AEGL program as it relates to the following:

- NAC/AEGL Committee and AEGL program objectives of scientific credibility, quality, productivity, and cost effectiveness.
- Resource needs of the AEGL program.
- Fostering of a collaborative spirit among committee members, staff scientists of the organization that drafts TSDs, and interested parties from all participating organizations in the public and private sectors.
- Matters related to the U.S. National Academy of Sciences–National Research Council.

- Expansion of the scope of the AEGL program, including international participation.

4.5 ROLE OF THE DESIGNATED FEDERAL OFFICER

The designated federal officer (DFO) serves as the administrative officer of the committee to ensure that all operations, processes, and general procedures function properly and efficiently. The DFO serves as an executive secretariat to the NAC/AEGL Committee and is responsible for the following:

- Effective communication and coordination with NAC/AEGL Committee members, the committee chair, the organization that drafts TSDs, and interested parties in the public and private sectors.
- Day-to-day administrative management of the NAC/AEGL Committee with respect to the agenda for future meetings, distribution of TSDs and other correspondence with committee members, maintenance of meeting minutes, tapes of meetings and other important committee records, funding and other financial matters, and committee membership matters.
- Administrative management of quarterly meetings, including responsibility for all *Federal Register* Notices related to NAC/AEGL Committee activities, minutes, decision-making records, meeting venues, facilities, and equipment, as well as the assurance that the meetings are held in compliance with the Federal Advisory Committee Act (FACA).
- Compliance with FACA on all matters that extend beyond the quarterly meetings, such as the submission of appropriate reports to the U.S. Office of Management and Budget (OMB) and the Library of Congress.

4.6 ROLE OF THE NAC/AEGL COMMITTEE CHAIR

The NAC/AEGL Committee chair is appointed by EPA as specified in FACA and selected from the NAC/AEGL Committee membership. The chair's responsibilities include conducting and directing specific activities to ensure the effective and efficient conduct of business by the committee:

- Provide support in the planning and preparation of upcoming meet-

ings by collaborating with the AEGL program director, the DFO, and the organization that drafts TSDs, including the review of the meeting agenda.
- Manage the NAC/AEGL Committee meetings in an effective and efficient manner to ensure completion of the agenda for each meeting.
- Attempt to reach a consensus of the NAC/AEGL Committee by ensuring adequate time for presentation of differing opinions and focusing on the major issues to break deadlocks or stalemates.
- Participate in scientific matters on AEGLs related to the U.S. National Academy of Sciences–National Research Council.
- Participate with the AEGL program director and the DFO in evaluating and improving NAC/AEGL Committee activities and expanding the scope of the AEGL program.

4.7 CLASSIFICATION OF THE STATUS OF AEGL VALUES

Draft AEGL values are AEGL values that have been proposed by the AEGL Development Team (see Section 4.8) before the full NAC/AEGL Committee discussion and approval.

Proposed AEGL values are AEGL values that have been formally approved and elevated to "proposed" status by a consensus or a two-thirds majority vote of the NAC/AEGL Committee.

Interim AEGL values are AEGL values that have been formally approved by the NAC/AEGL Committee and elevated to "interim" status after publication in the *Federal Register*, response to comments, and appropriate adjustments made by the committee. Interim AEGL values are forwarded to the Committee on Toxicology of the NRC for review and comment by the Subcommittee on Acute Exposure Guideline Levels (NRC/AEGL Subcommittee).

Final AEGL values are AEGL values that have been reviewed, finalized, and published by the NRC.

4.8 FUNCTION OF AEGL DEVELOPMENT TEAMS

Each AEGL Development Team consists of a staff scientist from the organization that drafts TSDs, a chemical manager, and two chemical reviewers who are members of the NAC/AEGL Committee. The primary function of the NAC/AEGL Development Team is to provide the NAC/AEGL Committee with draft AEGL values and a TSD containing relevant data and information on the chemical and the derivation of the draft AEGLs. The staff scientist

CURRENT ADMINISTRATIVE PROCESSES AND PROCEDURES   149

provides the initial effort by identifying and preliminarily evaluating available data from varied resources including on-line literature databases, other databases, journal reviews, secondary source reviews, unpublished data, federal and state documents and other sources, including accounts of accidents in the workplace or in the community (see Section 2.3). Interaction takes place among the chemical manager, the chemical reviewers, and the staff scientist during the development of the TSD and the draft AEGL values. The resulting document is then distributed and reviewed by committee members prior to a formal meeting, and attempts are made to resolve issues of concern expressed by committee members prior to distribution of the TSD to the NAC/AEGL Committee and formal presentation and discussion at a committee meeting.

4.8.1 Role of a Chemical Manager

The chemical manager has the overall responsibility for the development of the draft, proposed, and interim AEGL values and their presentation to the rest of the NAC/AEGL Committee and to the NRC/AEGL Subcommittee for evaluation of final AEGLs. The chemical managers serve on a rotating basis as the committee's principal representative on the AEGL Development Team for a specific chemical. The chemical manager in turn selects two committee members to serve as chemical reviewers.

The chemical manager collaborates with the staff scientist and the chemical reviewers on the development of the AEGLs, the supporting rationale, and the TSDs. In instances in which the chemical manager has accepted the responsibilities, taken ownership for the AEGL values, resolved scientific issues, and led the discussions with committee members, the NAC/AEGL Committee has moved rapidly toward the development of a consensus. Where the chemical manager's role has been less decisive, the committee's deliberations have been more protracted, less focused, and highly inefficient. Implicit in the description of the chemical manager's role is the expectation that he or she will work with the staff scientist, the chemical reviewers, and the rest of the committee members to develop exposure guidance levels that are appropriate and scientifically credible. It is expected that the chemical manager will achieve a consensus within the AEGL Development Team on the issues related to the development of the AEGL values prior to the meeting of the full NAC/AEGL Committee. Further, as time permits, the chemical manager will attempt to resolve issues raised by individual committee members prior to the scheduled committee meeting.

The following is a summary outline of specific activities and responsibilities of the chemical manager within the NAC/AEGL Committee:

- Participate as the leader of the ad hoc AEGL Development Team.
- Select and utilize two chemical reviewers as technical support.
- Provide direct support to the staff scientist assigned to the chemical in the development of the TSDs, the draft AEGL values, and the supporting rationale.
- Serve as liaison between NAC/AEGL Committee members and the staff scientist during the development of draft AEGL values and the TSD.
- Resolve scientific issues prior to the NAC/AEGL Committee meetings, such as the following:
  — Completeness of data gathering (published and unpublished).
  — Selection of key and supporting data (following guidelines).
  — Interpretation of data.
  — Credibility of AEGL values (use of appropriate methodology).
  — Validity of scientific rationale for AEGLs.
  — Other (as necessary for development of scientifically credible AEGL values).
- Seek consensus of NAC/AEGL Committee members by resolving issues with individual committee members prior to the committee meeting.
- Frame important scientific issues related to the chemical and the AEGLs for presentation at the NAC/AEGL Committee meeting (i.e., significant issues that cannot be resolved before the meeting).
- Participate in the presentation of AEGL values, supporting the rationale and important issues at the NAC/AEGL Committee meeting in collaboration with the staff scientist.
- Oversee appropriate follow-up activities:
  — Revisions as appropriate (AEGL values, TSD, rationales).
  — Toxicity testing.
  — *Federal Register* Notice comments (conversion of "proposed" to "interim" values).
  — Preparation of AEGL proposal to the NRC.

4.8.2 Role of a Chemical Reviewer

- Participate as a member of the ad hoc AEGL Development Team.
- Conduct a detailed review of the assigned document and key references.
- Assist the chemical manager and staff scientist in evaluating the data, the candidate AEGLs, and the scientific rationale for their support.

- Participate actively in discussions of the document during NAC/AEGL Committee meetings.
- Stand in for the chemical manager if he or she is unable to perform his or her duties.

4.8.3 Role of a Staff Scientist at the Organization That Drafts TSDs

The staff scientist has the primary responsibility for data gathering, data evaluation, identification of potential key data and supporting data, identification of potential methodologies, calculations, and extrapolations, and the preparation of the TSD. The following tasks are included:

- Participate as a member of the ad hoc AEGL Development Team.
- Participate with the others on the AEGL Development Team in the development of draft AEGL values and their presentation at the NAC/AEGL Committee meetings.
- Prepare TSDs in a timely manner and make appropriate revisions based on discussions and decisions of the AEGL Development Team and later based on the discussions and decisions of the NAC/AEGL Committee.
- Develop and maintain a data file on the chemical substance.
- Present a summary of the data and information on the substance in collaboration with the chemical manager at the NAC/AEGL Committee meetings.
- Provide continuing support to an assigned chemical through the draft, proposed, interim, and final stages of AEGL development, including preparation for, and response to, *Federal Register* Notice review and comment.

4.9 ROLE OF NAC/AEGL COMMITTEE MEMBERS

- Review all TSDs in advance of meetings, and work out issues with the chemical manager at the earliest possible date. The importance of resolving issues before NAC/AEGL Committee meetings is greatly emphasized to increase the efficiency and productivity of the meetings.
- Circulate TSDs to other qualified scientists within their respective organizations or other organizations as appropriate to broaden the evaluation by the scientific community.
- Serve as experts in specific areas or on specific scientific issues (e.g.,

sensitive human subpopulations) as a member of an ad hoc task force under the SOP Workgroup chair.
- Volunteer as a chemical manager at least once a year, and select chemicals on the basis of special knowledge, expertise, or past experience, whereby a significant contribution can be made to the development of credible AEGL values.
- Assist in the application of AEGLs in appropriate programs within the organization represented by the NAC/AEGL Committee member.
- Make suggestions for modification or expansion of the Chemical Priority List by providing lists of chemicals and supporting rationale for their priority to the designated federal officer (DFO).
- Attend all scheduled NAC/AEGL Committee meetings, and participate in the discussions and decision-making of all AEGL values. AEGL values are approved or disapproved by a two-thirds majority vote of the committee quorum (51% or more of the committee members present).

4.10 ROLE OF THE ORGANIZATION THAT DRAFTS TSDs

The role of the organization that drafts the TSDs is to provide the principal technical support in gathering and evaluating the relevant scientific data and information from all sources in the preparation and revision of the TSDs, following the guidance provided in this SOP manual. As a member of the AEGL Development Team,

- Collaborates with the chemical manager and chemical reviewers in the preparation and distribution of draft AEGLs, the supporting rationale, and the TSDs for the NAC/AEGL Committee members.
- Provides continuing technical and administrative support to assigned chemicals through the Draft, Proposed, Interim, and Final stages of AEGL development, with revisions based on the consensus or majority of the NAC/AEGL Committee and the NRC/AEGL Subcommittee.
- Provides the staff scientists and the administrative personnel with the facilities and equipment necessary for data gathering, maintenance of databases, dissemination of relevant information to committee members, presentations or co-presentations (with chemical managers) at the NAC/AEGL Committee meetings, development and revisions of TSDs, preparation of submissions to the *Federal Register*, summarization of *Federal Register* comments and identification of important scientific issues, presentations to the committee on *Federal Register*

comments, and preparation of technical information to be entered on the Internet.
- Distributes the TSDs to companies and other interested parties as directed by the DFO after review and comment by the NAC/AEGL Committee. Distribution to interested parties will be only by request through the DFO. The initial distributed version will be without the AEGL values and the rationale used to derive them and will occur between 1 and 14 days before the committee meeting.

References

AAAI (American Academy of Allergy and Immunology). 1986. Clinical ecology. Executive Committee of the American Academy of Allergy and Immunology. J. Allergy Clin. Immunol. 78(2):269-271.

AAAAI (American Academy of Allergy Asthma and Immunology). 1999. Physician Reference Materials: Position Statement 35: Idiopathic Environmental Tolerances. AAAAI Board of Directors. JACI 103:36-40. [Online]. Available: http://www.aaaai.org/professional/physicianreference/positionstatements/ps35.stm. [March 15, 1999].

Alder, H.L., and E.B. Roessler. 1968. P. 191 in Introduction to Probability and Statistics, 4th Ed. San Francisco: W.H. Freeman.

Aleksieva, Z. 1983. Sulphur compounds. Pp. 2122-2124 in Encyclopedia of Occupational Health and Safety, L. Prameggiani, ed. Geneva: International Labour Office.

Allen, B.C., R.J. Kavlock, C.A. Kimmel, and E.M. Faustman. 1994. Dose-response assessment for developmental toxicity. II. Comparison of generic benchmark dose estimates with no observed adverse effect levels. Fundam. Appl. Toxicol. 23(4):487-495.

Ashford, N. 1999. Low-level chemical sensitivity: Implications for research and social policy. Toxicol. Ind. Health 15(3-4):421-427.

Ashford, N., and C. Miller. 1998. Chemical Exposures: Low Levels and High Stakes, 2nd Ed. New York: John Wiley & Sons.

Balmes, J.R., J.M. Fine, and D. Sheppard. 1987. Symptomatic bronchoconstriction after short-term inhalation of sulfur dioxide. Am. Rev. Respir. Dis. 136(5):1117-1121.

Barnes, D.G., and M.L. Dourson. 1988. Reference dose (RfD): Description and use in health risk assessments. Regul. Toxicol. Pharmacol. 8(4):471-486.

Barnes, G.B., G.P. Daston, J.S. Evans, A.M. Jarabek, R.J. Kavlock, C.A. Kimmel, C.

Park, and H.L. Spitzer. 1995. Benchmark dose workshop: Criteria for use of a benchmark dose to estimate a reference dose. Regul. Toxicol. Pharmacol. 21(2):296-306.

Barnes, J.M., and F.A. Denz. 1954. Experimental methods used in determining chronic toxicity. Pharmacol. Rev. 6:191-242.

Bigwood, E.J. 1973. The acceptable daily intake of food additives. CRC Crit. Rev. Toxicol. 2(1):41-93.

Bliss, C.I. 1935a. The calculation of the dosage-mortality curve. Ann. Appl. Bio. 22:134-167.

Bliss, C.I. 1935b. The comparison of the dosage-mortality data. Ann. Appl. Bio. 22:307-333.

Bogdanffy, M.S., R. Sarangapani, D.R. Plowchalk, A. Jarabek, and M.E. Andersen. 1999. A biologically based risk assessment for vinyl acetate-induced cancer and noncancer inhalation toxicity. Toxicol. Sci. 51(1):19-35.

Bogen, K.T. 1989. Cell proliferation kinetic and multistage cancer risk models. J. Natl. Cancer Inst. 81(4):267-277.

Brown, N.A., and S. Fabro. 1983. The value of animal teratogenicity testing for predicting human risk. Clin. Obst. Gynecol. 26(2):467-477.

Brown-DeGagne, A.M., and J. McGlone. 1999. Multiple chemical sensitivity: A test of the olfactory-limbic model. J. Occup. Environ. Med. 41(5):366-377.

Bruckner, J.V., and W.B. Weil. 1999. Biological factors which may influence an older child's or adolescent's responses to toxic chemicals. Regul. Toxicol. Pharmacol. 29(2 Pt 1):158-164.

Calabrese, E.J. 1983. Principles of Animal Extrapolation. Reprinted 1991. New York: Wiley.

Calabrese, E.J. 1985. Uncertainty factors and interindividual variation. Regul. Toxicol. Pharmacol. 5:190-196.

Chan, M.T., P. Mainland, and T. Gin. 1996. Minimum alveolar concentration of halothane and enflurane are decreased in early pregnancy. Anesthesiology 85(4):782-786.

Chen, J.J., R.L. Kodell, and D.W. Gaylor. 1988. Using the biological two-stage model to assess risk from short-term exposures. Risk Anal. 8(2):223-230.

Crump, K.S., and R.B. Howe. 1984. The multistage model with a time-dependent dose pattern: Applications to carcinogenic risk assessment. Risk Anal. 4:163-176.

Cullen, M.R. 1987. The worker with multiple chemical sensitivities: An overview. Occup. Med. 2(4):655-661.

Done, A.K. 1964. Developmental pharmacology. Clin. Pharmacol. Ther. 5:432-479. (as cited in NRC, 1993b).

Dourson, M. 1996. Uncertainty factors in noncancer risk assessment. Regul. Toxicol. Pharmacol. 24(2):107.

Dourson, M.L., and J.F. Stara. 1983. Regulatory history and experimental support of uncertainty (safety) factors. Regul. Toxicol. Pharmacol. 3(3):224-238.

Dourson, M.L., S.P. Felter, and D. Robinson. 1996. Evolution of science-based uncertainty factors in noncancer risk assessment. Regul. Toxicol. Pharmacol. 24(2):108-120.

Dourson, M.L., L.A. Knauf, and J.C. Swartout. 1992. On reference dose (RfD) and its underlying toxicity data base. Toxicol. Ind. Health 8(3):171-189.

ECETOC (European Chemical Industry Ecology and Toxicology Centre). 1991. Pp. 28-29 in Emergency Exposure Indices for Industrial Chemicals. Tech. Report no. 43. Brussels, Belgium: European Chemical Industry Ecology and Toxicology Centre.

EPA (U.S. Environmental Protection Agency). 1980. Guidelines and Methodology Used in the Preparation of Health Assessment Chapters of the Consent Decree Water Quality Criteria. Fed. Regist. 45:79347-79357.

EPA (U.S. Environmental Protection Agency). 1986. Guidelines for Carcinogen Risk Assessment. Fed. Regist. 51(185):33992-34003.

EPA (U.S. Environmental Protection Agency). 1987. Technical Guidance for Hazards Analysis. Emergency Planning for Extremely Hazardous Substances. U.S. Environmental Protection Agency, Federal Emergency Management Agency, U.S. Department of Transportation, Washington, DC.

EPA (U.S. Environmental Protection Agency). 1991. National Primary Drinking Water Regulations, Final rule. 40 CFR Parts 141, 142, and 143. Fed. Regist. 56(20):3526-3597. January 30.

EPA (U.S. Environmental Protection Agency). 1992. Draft report: A Cross-Species Scaling Factor for Carcinogen Risk Assessment Based on Equivalence of $mg/kg^{3/4}/day$. Fed. Regist. 57(109):24152-24173.

EPA (U.S. Environmental Protection Agency). 1993. Styrene. The Integrated Risk Information System (IRIS). National Center for Environmental Assessment, Washington, D.C. Available: http://www.epa.gov/ngispgm3/iris/subst/0104.htm [1/3/01, last updated 5/5/98].

EPA (U.S. Environmental Protection Agency). 1994. Methods for Derivation of Inhalation Reference Concentrations and Application of Inhalation Dosimetry. EPA/600/8-90/066F. Environmental Criteria and Assessment Office, Office of Research and Development, U.S. Environmental Protection Agency, Research Triangle Park, NC. October.

EPA (U.S. Environmental Protection Agency). 1995a. The Use of the Benchmark Dose Approach in Health Risk Assessment. EPA/630/R-94/007. Risk Assessment Forum. U.S. Environmental Protection Agency, Washington, DC. February.

EPA (U.S. Environmental Protection Agency). 1995b. Methylmercury. The Integrated Risk Information System (IRIS). National Center for Environmental Assessment, Washington, D.C. Available: http://www.epa.gov/ngispgm3/iris/subst/0073.htm/. [1/3/01, last updated 5/5/98].

EPA (U.S. Environmental Protection Agency). 1996a. Proposed Guidelines for Carcinogen Risk Assessment. EPA/600/P-92/003C. Office of Research and Development, U.S. Environmental Protection Agency, Washington, DC. April.

EPA (U.S. Environmental Protection Agency). 1996b. Aroclor 1016. The Integrated Risk Information System (IRIS). National Center for Environmental Assessment, Washington, D.C. Available: http://www.epa.gov/iris/.

EPA (U.S. Environmental Protection Agency). 1999. Health Effects Test Guidelines. Code of Federal Regulations 40: Part 798. Available: http://www.epa.gov/docs/OPPTS_Harmonized/.

EPA (U.S. Environmental Protection Agency). 2000. Benchmark Dose Software Draft Beta Version 1.2., National Center for Environmental Assessment, Office of Research and Development. [Online]. Available: http://www.epa.gov/ncea/bmds.htm. [1/3/01, last updated 1/2/01].

Evans, R.D., R.S. Harris, and J.W.M. Bunker. 1944. Radium metabolism in rats, and the production of osteogenic sarcoma by experimental radium poisoning. Am. J. Roentgenol. 52:353-373.

Faustman, E.M., B.C. Allen, R.J. Kavlick, and C.A. Kimmel. 1994. Dose-response assessment for developmental toxicity. I. Characterization of database and determination of no observed adverse effect levels. Fundam. Appl. Toxicol. 23(4):478-486.

FDA (U.S. Food and Drug Administration). 1985. Sponsored compounds in food-producing animals. Proposed Rule and Notice. Fed. Regist. 50:45530-45556.

Finney, D.J. 1971. Probit Analysis, 3rd Ed. London: Cambridge University Press.

Flury, F. 1921. Über Kampfgasvergiftungen. I. Über Reizgase. Z. Ges. Exp. Med. 13:1.

Food Safety Council. 1982. A Proposed Food Safety Evaluation Process. The Nutrition Foundation, Inc., Washington, DC.

Fowles, J.R., G.V. Alexeeff, and D. Dodge. 1999. The use of benchmark dose methodology with acute inhalation lethality data. Regul. Toxicol. Pharmacol. 29(3):262-278.

Frank, R. 1980. SO_2-particulate interactions: recent observations. Am. J. Ind. Med. 1(3-4):427-434.

Frederick, C.B., M.L. Bush, L.G. Lomax, K.A. Black, L. Finch, J.S. Kimbell, K.T. Morgan, R.P. Subramaniam, J.B. Morris, and J.S. Ultman. 1998. Application of a hybrid computational fluid dynamics and physiologically based inhalation model for interspecies dosimetry extrapolation of acidic vapors in the upper airways. Toxicol. Appl. Pharmacol. 152(1):211-231.

Gaylor, D.W. 1996. Quantalization of continuous data for benchmark dose estimation. Regul. Toxicol. Pharmacol. 24(3):246-250.

Gaylor, D.W., R.L. Kodell, J.J. Chen, and D. Krewski. 1999. A unified approach to risk assessment for cancer and noncancer endpoints based on benchmark doses and uncertainty/safety factors. Regul. Toxicol. Pharmacol. 29(2 Pt 1):151-157.

Gaylor, D.W., L. Ryan, D. Krewski, and Y. Zhu. 1998. Procedures for calculating benchmark doses for health risk assessment. Regul. Toxicol. Pharmacol. 28(2):150-164.

Glaubiger, D.L., D.D. von Hoff, J.S. Holcenbern, B. Kamen, C. Pratt, and R.S. Ungerleider. 1982. The relative tolerance of children and adults to anticancer drugs. Front. Radiat. Ther. Oncol. 16:42-49.

Goddard, M.J., M.J. Murdoch, and D. Krewski. 1995. Temporal aspects of risk characterization. Inhalation Toxicol. 7(6):1005-1018.

Goldenthal, E.I. 1971. A compilation of LD_{50} values in newborn and adult animals. Toxicol. Appl. Pharmacol. 18(1):185-207.

Graveling, R.A., A. Pilkington, J.P. George, M.P. Butler, and S.N. Tannahill. 1999. A review of multiple chemical sensitivity. Occup. Environ. Med. 56(2):73-85.

Gregory, G.A., E.I. Eger 2nd, and E.S. Munson. 1969. The relationship between age and halothane requirements in man. Anesthesiology 30(5):488-491.

Guth, D.J., A.M. Jarabek, L. Wymer, and R.C. Hertzberg. 1991. Evaluation of Risk Assessment Methods for Short-Term Inhalation Exposure. Paper No. 91-173.2. Paper presented at the 84th Annual Meeting of the Air and Waste Management Association, June 16-21, 1991, Vancouver, BC.

Haber, F. 1924. Zur Geschichte des Gaskrieges. Pp. 76-92 in Fünf Vorträge aus den Jahren 1920-1923. Berlin: Springer-Verlag.

Hattis, D., and E. Anderson. 1999. What should be the implications of uncertainty, variability, and inherent "biases"/"conservatism" for risk management decision-making? Risk Anal. 19(1):95-107.

Hattis, D., and K. Barlow.. 1996. Human interindividual variability in cancer risks–technical and management challenges. Human Ecol. Risk Assesses. 2(1):194-220.

Hattis, D., and D.E. Burmaster. 1994. Assessment of variability and uncertainty distributions for practical risk analyses. Risk Anal. 14(5):713-730.

Hattis, D., P. Banati, and R. Goble 1999. Distributions of individual susceptibility among humans for toxic effects—For what fraction of which kinds of chemicals and effects does the traditional 10-fold factor provide how much protection? Ann. N.Y. Acad. Sci. 895:286-316.

Hattis, D., L. Erdreich, and M. Ballew. 1987. Human variability in susceptibility to toxic chemicals: A preliminary analysis of pharmacokinetic data from normal volunteers. Risk Anal. 7(4):415-426.

Hayes, W.J. 1967. Toxicity of pesticides to man: Risks from present levels. Proc. R. Soc. Lond. B. Biol. Sci. 167(7):101-127.

Henderson, R. 1992. Short-term exposure guidelines for emergency response: The approach of the Committee on Toxicology. Pp. 89-92 in Conference on Chemical Risk Assessment in the Department of Defense (DOD): Science Policy and Practice, H.J. Clewell, ed. Cincinnati, OH: American Conference of Governmental Industrial Hygienists.

Hertzberg, R.C., and M. Miller. 1988. A statistical model for species extrapolation using categorical response data. Toxicol. Ind. Health 1(4):43-57.

Hertzberg, R.C., and L. Wymer. 1991. Modeling the severity of toxic effects. Paper no. 91-173.4. Paper presented at the 84th Annual Meeting of the Air and Waste Management Association, June 16-21, 1991, Vancouver, BC.

Horstman, D.H., E. Seal, L.J. Folinsbee, P. Ives, and J. Roger. 1988. The relationship between exposure duration and sulfur dioxide-induced bronchoconstriction in asthmatic subjects. Am. Ind. Hyg. Assoc. J. 49:38-47.

Howe, R.B., K.S. Crump, and C. Van Landingham. 1986. GLOBAL86: A Computer Program to Extrapolate Quantal Animal Toxicity Data to Low Doses. Subcontract No. 2-251U-2745. Prepared for U.S. Environmental Protection Agency, Washington, DC.

Interagency Workgroup on Multiple Chemical Sensitivity. 1998. A Report on Multiple Chemical Sensitivity (MCS). A predecisional draft. The Environmental Health Policy Committee. [Online]. Available: http://web.health.gov/environment/mcs/toc.htm. [August 24, 1998].

REFERENCES

IPCS (International Programme on Chemical Safety). 1994. Assessing Human Health Risks of Chemicals: Derivation of Guidance Values for Health-Based Exposure Limits. Environmental Health Criteria No. 170. Geneva: World Health Organization.

IPCS (International Programme on Chemical Safety). 1996. Report of Multiple Chemical Sensitivities Workshop. February 21-23, 1996. Berlin, Germany.

Jarabek, A.M., M.G. Menache, J.H. Overton, M.L. Dourson, and F.J. Miller. 1990. The U.S. Environmental Protection Agency's inhalation RfD methodology: Risk assessment for air toxics. Toxicol. Ind. Health 6(5):279-302.

Katoh, T., and K. Ikeda. 1992. Minimum alveolar concentration of sevoflurane in children. Br. J. Anaesth. 68(2):139-141.

Kepler, G.M., R.B. Richardson, K.T. Morgan, and J.S. Kimbell. 1998. Computer simulation of inspiratory nasal airflow and inhaled gas uptake in a rhesus monkey. Toxicol. Appl. Pharmacol. 150(1):1-11.

Keplinger, M.L., and L.W. Suissa. 1968. Toxicity of fluorine short-term inhalation. Am. Ind. Hyg. Assoc. J. 29(1):10-18.

Kimbell, J.S., E.A. Gross, D.R. Joyner, M.N. Godo, and K.T. Morgan. 1993. Application of computational fluid dynamics to regional dosimetry of inhaled chemicals in the upper respiratory tract of the rat. Toxicol. Appl. Pharmacol. 121(2):253-263.

Kimbell, J.S., E.A. Gross, R.B. Richardson, R.B. Conolly, and K.T. Morgan. 1997a. Correlation of regional formaldehyde flux predictions with the distribution of formaldehyde-induced squamous metaplasia in F344 rat nasal passages. Mutat. Res. 380(1-2):143-154.

Kimbell, J.S., M.N. Godo, E.A. Gross, D.R. Joyner, R.B. Richardson, and K.T. Morgan. 1997b. Computer simulation of inspiratory airflow in all regions of the F344 rat nasal passages. Toxicol. Appl. Pharmacol. 145(2):388-398.

Kipen, H.M., and N. Fiedler. 1999. Invited commentary: Sensitivities to chemicals-context and implications. Am. J. Epidemiol. 150(1):13-16.

Klaassen, C.D., and J. Doull. 1980. Evaluation of safety: Toxicological evaluation. Pp. 26 in Cassarett and Doull's Toxicology, The Basic Science of Poisons, J. Doull, C.D. Klaassen, and M.O. Amdur, eds. New York.: Macmillan.

Kodell, R.L., D.W. Gaylor, and J.J. Chen. 1987. Using average lifetime dose rate for intermittent exposures to carcinogens. Risk Anal 7(3):339-345.

Koenig, J.Q., W.E. Pierson, M. Horike, and R. Frank. 1981. Effects of SO_2 plus NaCl aerosol combined with moderate exercise on pulmonary function in asthmatic adolescents. Environ. Res. 25(2):340-348.

Koenig, J.Q., W.E. Pierson, M. Horike, and R. Frank. 1982. Effects of inhaled sulfur dioxide (SO_2) on pulmonary function in healthy adolescents: Exposure to SO_2 + sodium chloride droplet aerosol during rest and exercise. Arch. Environ. Health. 37(1):5-9.

Koenig, J.Q., K. Dumler, V. Rebolledo, P.V. Williams, and W.E. Pierson. 1993. Respiratory effects of inhaled sulfuric acid on senior asthmatics and nonasthmatics. Arch. Environ. Health 48(3):171-175.

Kokoski, C.J. 1976. Written testimony of Charles J. Kokoski, Docket No. 76N-0070. Food and Drug Administration, Washington, DC.

Kreutzer, R., R.R. Neutra, and N. Lashuay. 1999. Prevalence of people reporting sensitivities to chemicals in a population-based survey. Am. J. Epidemiol. 150(1):1-12.

LeDez, K.M., and J. Lerman. 1987. The minimum alveolar concentration (MAC) of isoflurane in preterm neonates. Anesthesiology 67(3):301-307.

Lehman, A.J., and O.G. Fitzhugh. 1954. 100-fold margin of safety. Assoc. Food Drug Off. U.S.Q. Bull. 18:33-35.

Leisenring, W. and L. Ryan. 1992. Statistical properties of the NOAEL. Regul. Toxicol. Pharmacol. 15(2 Pt 1):161-171.

Lerman, J., S. Robinson, M.M. Willis, and G.A. Gregory. 1983. Anesthetic requirements for halothane in young children 0-1 month and 1-6 months of age. Anesthesiology 59(5):421-424.

Linn, W.S., E.L. Avol, R.C. Peng, D.A. Shamoo, and J.D. Hackney. 1987. Replicated dose-response study of sulfur dioxide effects in normal, atopic, and asthmatic volunteers. Am. Rev. Respir. Dis. 136(5):1127-1134.

Litchfield Jr., J.T., and F. Wilcoxon. 1948. A simplified method of evaluating dose-effect experiments. J. Pharmacol. Exp. Ther. 96:99-113.

Lu, F.C. 1979. Assessments at an international level of health hazards to man of chemicals shown to be carcinogenic in laboratory animals. Pp. 315-328 in Regulatory Aspects of Carcinogenesis, F. Coulston, ed. New York: Academic Press.

Lu, F.C. 1988. Acceptable daily intake: Inception, evolution, and application. Regul. Toxicol. Pharmacol. 8(1):45-60.

Lu, F.C., and R.L. Sielken, Jr. 1991. Assessment of safety/risk of chemicals: Inception and evolution of the ADI and dose-response modeling procedures. Toxicol. Lett. 59(1-3):5-40.

Marsoni, S., R.S. Ungerleider, S.B. Hurson, R.M. Simon, and L.D. Hammershaimb. 1985. Tolerance to antineoplastic agents in children and adults. Cancer Treat. Rep. 69(11):1263-1269.

McLellan, R.K., C.E. Becker, J.B. Borak, C. Coplein, A.M. Ducatman, E.A. Favata, J.F. Green, J. Herzstein, A.T. Jolly, J. Kalnas, K. Kulig, H.M. Kipen, D.C. Logan, F.L. Mitchell, H.W. McKinnon, M.A. Roberts, M. Russi, H.J. Sawyer, M.J. Sepulveda, M.J. Upfal, and M.C. Zepeda. 1999. ACOEM Position statement. Multiple chemical sensitivities: Idiopathic environmental intolerance. J. Occup. Environ. Med. 41(11):940-942.

Meek, M.E., R. Newhook, R.G. Liteplo, and V.C. Armstrong. 1994. Approach to assessment of risk to human health for priority substances under the Canadian Environmental Protection Act. Environ. Carcinogen. Ecotoxicol. Rev. 12(2):105-134.

Mitchell, F.L. 1995. Multiple Chemical Sensitivity: A Scientific Overview. Princeton, N.J.: Princeton Scientific.

Moolgavkar, S.H., and G. Luebeck. 1990. Two-event model for carcinogenesis: Biological, mathematical, and statistical considerations. Risk Anal. 10(2):323-341.

Morrison, P.F. 1987. Effects of time-variant exposure on toxic substance response. Environ. Health Perspect. 76:133-140.

Murdoch, D.J., and D. Krewski. 1988. Carcinogenic risk assessment with time-dependent exposure patterns. Risk Anal. 8(4):521-530.

Murdoch, D.J., D. Krewski, and J. Wargo. 1992. Cancer risk assessment with intermittent exposure. Risk Anal. 12(4):569-577.
NRC (National Research Council). 1977. Drinking Water and Health. Washington, DC: National Academy Press.
NRC (National Research Council). 1983. Risk Assessment in the Federal Government. Washington, DC: National Academy Press.
NRC (National Research Council). 1985. Emergency and Continuous Exposure Guidance Levels for Selected Airborne Contaminants, Vol. 5. Washington, DC: National Academy Press.
NRC (National Research Council). 1986. Criteria and Methods for Preparing Emergency Exposure Guidance Level (EEGL), Short-Term Public Emergency Guidance Level (SPEGL), and Continuous Exposure Guidance Level (CEGL) Documents. Appendix F Washington, DC: National Academy Press.
NRC (National Research Council). 1992a. Guidelines for Developing Spacecraft Maximum Allowable Concentrations for Space Station Contaminants. Washington, DC: National Academy Press.
NRC (National Research Council). 1992b. Multiple Chemical Sensitivities: A Workshop. Washington, DC: National Academy Press.
NRC (National Research Council). 1992c. Biologic Markers in Immunotoxicology. Washington, DC: National Academy Press.
NRC (National Research Council). 1993a. Guidelines for Developing Community Emergency Exposure Levels for Hazardous Substances. Washington, DC: National Academy Press.
NRC (National Research Council). 1993b. Pesticides in the Diets of Infants and Children. Washington, DC: National Academy Press.
NRC (National Research Council). 1994. Science and Judgment in Risk Assessment. Washington, DC: National Academy Press.
NRC (National Research Council). 2000. Acute Exposure Guideline Levels for Selected Airborne Chemicals, Vol.1. Washington, DC: National Academy Press.
OECD (Organization for Economic Cooperation and Development). 1993. Guidelines for the Testing of Chemicals, ed 1993 11th Addendum, 2000, Paris: OECD. Available: http://www.oecd.org/ehs/test/testlist.htm.
PCCRARM (U.S. Presidential/Congressional Commission on Risk Assessment and Risk Management). 1997. Risk Assessment and Risk Management in Regulatory Decision-Making. Final Report, Vol 2. Washington, DC: U.S. Government Printing Office.
Pieters, M.N., and H.J. Kramer. 1994. Concentration · Time = Constant ? The Validity of Haber's Law in the Extrapolation of Discontinuous to Continuous Exposition. Rapportnummer 659101 002, National Institute for Public Health and Environmental Protection, The Netherlands.
Pohl, H.R., and H.G. Abadin. 1995. Utilizing uncertainty factors in minimal risk levels derivation. Regul. Toxicol. Pharmacol. 22(2):180-188.
Renwick, A.G. 1993. Data derived safety factors for the evaluation of food additives and environmental contaminants. Food Addit. Contam. 10(3):275-305.
Rhomberg, L.R., and S.K. Wolff. 1998. Empirical scaling of single oral lethal doses across mammalian species based on a large database. Risk Anal. 18(6):741-753.

Rinehart, W.E., and T. Hatch. 1964. Concentration-time product (CT) as an expression of dose in sublethal exposures to phosgene. Ind. Hyg. J. 25:545-553.

Rondinelli, R.C., J.Q. Koenig, and S.G. Marshall. 1987. The effects of sulfur dioxide on pulmonary function in healthy nonsmoking male subjects aged 55 years and older. Am. Ind. Hyg. Assoc. J. 48(4):299-303.

Roger, L.J., H.R. Kehrl, M. Hazucha, and D.H. Horstman. 1985. Bronchoconstriction in asthmatics exposed to sulfur dioxide during repeated exercise. J. Appl. Physiol. 59(3):784-791.

Schlesinger, R.B. and I. Jaspers. 1997. Sulfur oxide. Pp 313-330 in Comprehensive Toxicology, Vol.8. Toxicology of Respiratory System, R. Roth, ed. New York: Pergamon.

Schlesinger, R.B. 1999. Nitrogen oxides. Pp. 595-638 in Environmental Toxicants, 2nd. Ed., M. Lippmann, ed. New York; Chichester: Wiley.

Sheenan, D.M., and D.W. Gaylor. 1990. Divisions of reproductive & developmental toxicology and biometry. [Abstract]. Teratology 41:590-591.

Simon, R.A. 1986. Sulfite sensitivity. Ann. Allergy 56(4):281-288.

Spiegel, M.R. 1996. Schaum's Outline of Theory and Problems of Statistics, 2nd Ed. New York; London: McGraw-Hill. 519 pp.

Stevens, W.D., W.M. Dolan, R.T. Gibbons, A. White, E.I. Eger, R.D. Miller, R.H. deJong, and R.M. Elashoff. 1975. Minimum alveolar concentration (MAC) of isoflurande with and without nitrous oxide in patients of various ages. Anesthesiology 42(2):197-200.

ten Berge, W.F., A. Zwart, and L.M. Appelman. 1986. Concentration-time mortality response relationship of irritant and systemically acting vapours and gases. J. Hazard. Mater. 13(3):301-309.

Truhaut, R. 1991. The concept of the acceptable daily intake: An historical review. Food Addit. Contam. 8(2):151-162.

Vettorazzi, G. 1977. Safety factors and their application in the toxicological evaluation. Pp. 207-223 in The Evaluation of Toxicological Data for the Protection of Public Health: Proceedings of the International Colloquium, Luxembourg, December 1976. Oxford: Pergamon.

Vettorazzi, G. 1980. Pp. 66-68 in Handbook of International Food Regulatory Toxicology. Vol.1. Evaluations. New York: SP Medical & Scientific Books.

Vocci, F., and T. Farber. 1988. Extrapolation of animal toxicity data to man. Regul. Toxicol. Pharmacol. 8(4):389-398.

Weeks, M.H., G.C. Maxey, M.E. Sicks, and E.A. Greene. 1963. Vapor toxicity of UDMH in rats and dogs from short exposures. Am. Ind. Hyg. Assoc. J. 24(2):137-143.

Weil, C. 1972. Statistics vs safety factors and scientific judgment in the evaluation of safety for man. Toxicol. Appl. Pharmacol. 21(4):454-463.

Appendixes

Appendix A

Priority Lists of Chemicals

A master list of approximately 1,000 acutely toxic chemicals was initially compiled through the integration of individual priority lists of chemicals submitted by each U.S. federal agency placing a representative on the NAC/AEGL Committee. The master list was subsequently reviewed by individuals from certain state agencies and representatives from organizations in the private sector and modified as a result of comments and suggestions received. The various priority chemical lists were compiled separately by each federal agency based on their individual assessments of the hazards, potential exposure risk, and relevance of a chemical to their programmatic needs.

On May 21, 1997, a list of 85 chemicals was published in the *Federal Register*. This list identified those chemicals to be of highest priority across all U.S. federal agencies and represented the selection of chemicals for AEGL development by the NAC/AEGL Committee for the first 2-3 years of the program. The committee has now addressed most of these chemicals, and they are currently in the "proposed," "interim," or "final" stages of development. Certain chemicals did not contain an adequate database for AEGL development and, consequently, are on hold pending decisions regarding further toxicity testing. This initial "highest" priority list of 85 chemicals is shown in Table A-1.

A second "working list" of approximately 100 priority chemicals is being selected from the original master list or from new high-priority candidate chemicals submitted by U.S. agencies and organizations and by member countries of the Organization for Economic Cooperation and Development (OECD) that are planning to participate in the AEGL Program. Although "working lists" will be published in the *Federal Register* and elsewhere from time to time to indicate the NAC/AEGL Committee's agenda, the priority of chemicals addressed, and, hence, the "working list," is subject to modification if priorities of the NAC/AEGL Committee or individual stakeholder organizations, including international members, change during that period.

INITIAL LIST OF 85 PRIORITY CHEMICALS FOR AEGL DEVELOPMENT

Organization Lists Used to Compile the Master List and The Initial List of 85 Priority Chemicals[1]

ATSDR Medical Managment U.S. Agency for Toxic Substances and Disease Registry
    M = Chemicals with an ATSDR Medical Management Guideline
    T = Chemicals with an ATSDR Toxicology Profile

DOD U.S. Department of Defense
    A = Army Toxicity Summary Chemical
    C = Chemical Weapons Convention Schedule 3.A Toxic Chemical
    Cs = Chemical Stockpile Emergency Preparedness Program (CSEPP) Chemical

---

[1] The initial list of 85 priority chemicals shown in Table A-1 has been created by identifying the highest priority hazardous chemicals from the Master List. This initial list is a starting point for the development of AEGL values by the National Advisory Committee for Acute Exposure Guideline Levels for Hazardous Chemicals (NAC/AEGL). However, the list of chemicals is subject to modification, pending changes in priorities recommended by the various stakeholders that make up the NAC/AEGL. While it is anticipated that most of these chemicals will remain as high priority for AEGL development, changes to the list could occur. The NAC/AEGL Committee hopes to select 30 to 40 chemicals per year to address in the AEGL development process. Consequently, the initial list will expand as the NAC/AEGL Committee continues to address chemicals of interest to its member organizations.

APPENDIX A

	I = Air Force Installation Restoration Program Chemical
	N = Navy Chemical
	S = Strategic Environmental Research and Development Program (SERDP) Chemical
DOE SCAPA	U.S. Department of Energy Subcommittee for Consequence Assessment and Protective Action Chemical
DOT ERP	U.S. Department of Transportation Emergency Response Guidebook
	P = Priority DOT ERG Chemical
	O = Other ERG Chemical
EPA CAA 112b	U.S. Environmental Protection Agency Clean Air Act 112b Chemical
EPA CAA 112r	U.S. Environmental Protection Agency Clean Air Act 112b Chemical (+ = SARA s.302 also)
EPA Superfund	U.S. Environmental Protection Agency Superfund Chemical
OSHA PSM	U.S. Occupational Safety and Health Administration Process Safety Management Chemical
OSHA STEL	U.S. Occupational Safety and Health Administration Short-term Exposure Limit Chemical
NIOSH IDLH	NIOSH Immediately Dangerous to Life or Health Chemical
Seveso Annex III	International Seveso Convention List

TABLE A-1 Priority List of Chemicals

CAS No.	Chemical	ATSDR	DOD	DOE SCAPA	DOT ERG	EPA CAA 112b	EPA CAA 112r	EPA Superfund	OSHA PSM	OSHA STEL	NIOSH IDLH	Seveso Annex III
56-23-5	Carbon tetrachloride	T	AIS			X		X			X	
57-14-7	1,1-Dimethyl hydrazine				P	X	X+		X		X	
60-34-4	Methyl hydrazine				P	X	X+	X	X		X	
62-53-3	Aniline	M			P	X	+	X			X	
67-66-3	Chloroform	T	AIS			X	X+	X			X	
68-12-2	Dimethylformamide			X		X						
71-43-2	Benzene	X	AIS	X		X		X				
71-55-6	1,1,1-Trichloroethane	T	X	X		X		X				
74-90-8	Hydrogen cyanide	M	C		P	X	X+		X		X	X
74-93-1	Methyl mercaptan	T			P		X+		X		X	
75-09-2	Methylene chloride	MT	AIS	X		X		X				
75-21-8	Ethylene oxide	MT			P	X	X+		X		X	X
75-44-5	Phosgene	M	C		P	X	X+		X		X	X
75-55-8	Propyleneimine					X	X+				X	X
75-56-9	Propylene oxide					X	X+				X	X
75-74-1	Tetramethyllead					X	X+		X		X	X
75-77-4	Trimethychlorosilane						X+					
75-78-5	Dimethyldichlorosilane			X			X+		X			
75-79-6	Methyltrichlorosilane						X+		X			
78-82-0	Isobutyronitrile						X+					
79-01-6	Trichloroethylene	MT	AIS	X		X		X				

CAS #	Name										
79-21-0	Peracetic acid					X+					X
79-22-1	Methy chloroformate					X+					
91-08-7	Toluene 2,6-diisocyanate	M				X+		X			
106-89-8	Epichlorohydrin				X	X+				X	
107-02-8	Acrolein	T		P	X	X+	X	X	X	X	X
107-11-9	Allyl amine			P		X+		X		X	X
107-12-0	Propionitrile					X+					
107-15-3	Ethylenediamine					X+				X	
107-18-6	Allyl alcohol			P		X+			X	X	X
107-30-2	Chloromethyl methyl ether			O	X	X+		X			X
108-23-6	Isopropyl chloroformate			P		X+					
108-88-3	Toluene	MT	AINS				X				
108-91-8	Cyclohexylamine					X+					
109-61-5	Propyl chloroformate			O		X+		X			
110-00-9	Furan					X+	X				
110-89-4	Piperidine					X+					
123-73-9	Crotonaldehyde, (E)					X+	X			X	
126-98-7	Methacrylonitrile			O	X	X+		X			
127-18-4	Tetrachloroethylene	T	AIS		X		X				
151-56-4	Ethyleneimine			P	X	X+			X	X	
302-01-2	Hydrazine	T	I		X	X+				X	X
353-42-4	Boron triflouride compound with methyl ether (1:1)					X+				X	
506-77-7	Cyanogen chloride							X			X
509-14-8	Tetranitromethane					X+			X		

TABLE A-1 (Continued)

CAS No.	Chemical	ATSDR	DOD	DOE SCAPA	DOT ERG	EPA CAA 112b	EPA CAA 112r	EPA Superfund	OSHA PSM	OSHA STEL	NIOSH IDLH	Seveso Annex III
540-59-0	1,2-Dichloroethylene	T		X							X	
540-73-8	1,2-Dimethylhydrazine				P	X	X+		X		X	
584-84-9	Toluene 2,4-diisocyanate	M				X	X+	X		X	X	
594-42-3	Perchloromethyl-mercaptan						X+		X		X	X
624-83-9	Methyl isocyanate				P	X	X+		X		X	X
811-97-2	HFC 134A (1,1,1,2-Tetrafluoroethane)		N									
814-68-6	Acrylyl chloride						X+		X			
1330-20-7	Xylenes (mixed)	X	AIN			X		X				
1717-00-6	HCFC 141b (1,1-Dichloro-1-fluoroethane)		N									
4170-30-3	Crotonaldehyde cis & trans mixture				P		X+				X	
6423-43-4	Propylene glycol dinitrate (Otto Fuel II)	T	Navy									
7446-09-5	Sulfur dioxide				P		X+		X	X		X
7446-11-9	Sulfur trioxide				P		X+		X			X
7647-01-0	Hydrogen chloride				P	X	X+	X	X	X	X	X
7647-01-0	Hydrochloric acid				P	X	X+	X	X	X	X	
7664-39-3	Hydrogen fluoride	M			P	X	X+		X	X	X	X
7664-41-7	Ammonia	MT					X+	X	X	X	X	X
7664-93-9	Sulfuric acid				P		+	X			X	
7697-37-2	Nitric acid			X	P		X+		X	X	X	

171

CAS #	Name									
7719-12-2	Phosphorus trichloride							X	X	
7726-95-6	Bromine						X	X	X	X
7782-41-4	Fluorine						X	X	X	
7782-50-5	Chlorine	M		P	X+		X	X	X	X
7783-06-4	Hydrogen sulfide	M		X	X+	X	X			
7783-60-0	Sulfur tetrafluoride			P	X+					
7783-81-5	Uranium hexafluoride		X							
7784-34-1	Arsenous trichloride			P	X+					
7784-42-1	Arsine	M	X	P	X+	X	X		X	X
7790-91-2	Chlorine trifluoride		X	O			X		X	
7803-51-2	Phosphine	M	X	P	X+		X	X	X	X
8014-95-7	Oleum			P	X+		X			
10025-87-3	Phosphorus oxychloride			O	X+		X			
10049-04-4	Chlorine dioxide				X		X	X	X	
10102-43-9	Nitric oxide			P	X+		X		X	
10102-44-0	Nitrogen dioxide			X	X		X			
10294-34-5	Boron trichloride			P	X+		X			
13463-39-3	Nickel carbonyl			P	X	X+	X	X		X
13463-40-6	Iron, pentacarbonyl-			P	X+		X	X		
19287-45-7	Diborane		X	P			X		X	
25323-89-1	Trichloroethane	T	X		X	X				
70892-10-3	Jet fuels (JP-5 and JP-8)	N								
163702-07-6	Methyl nonafluorobutyl ether (HFE 7100 component)	N								
163702-08-7	Methyl nonafluorobutyl ether (HFE 7100 component)	N								

Appendix B

Diagram of the AEGL Development Process

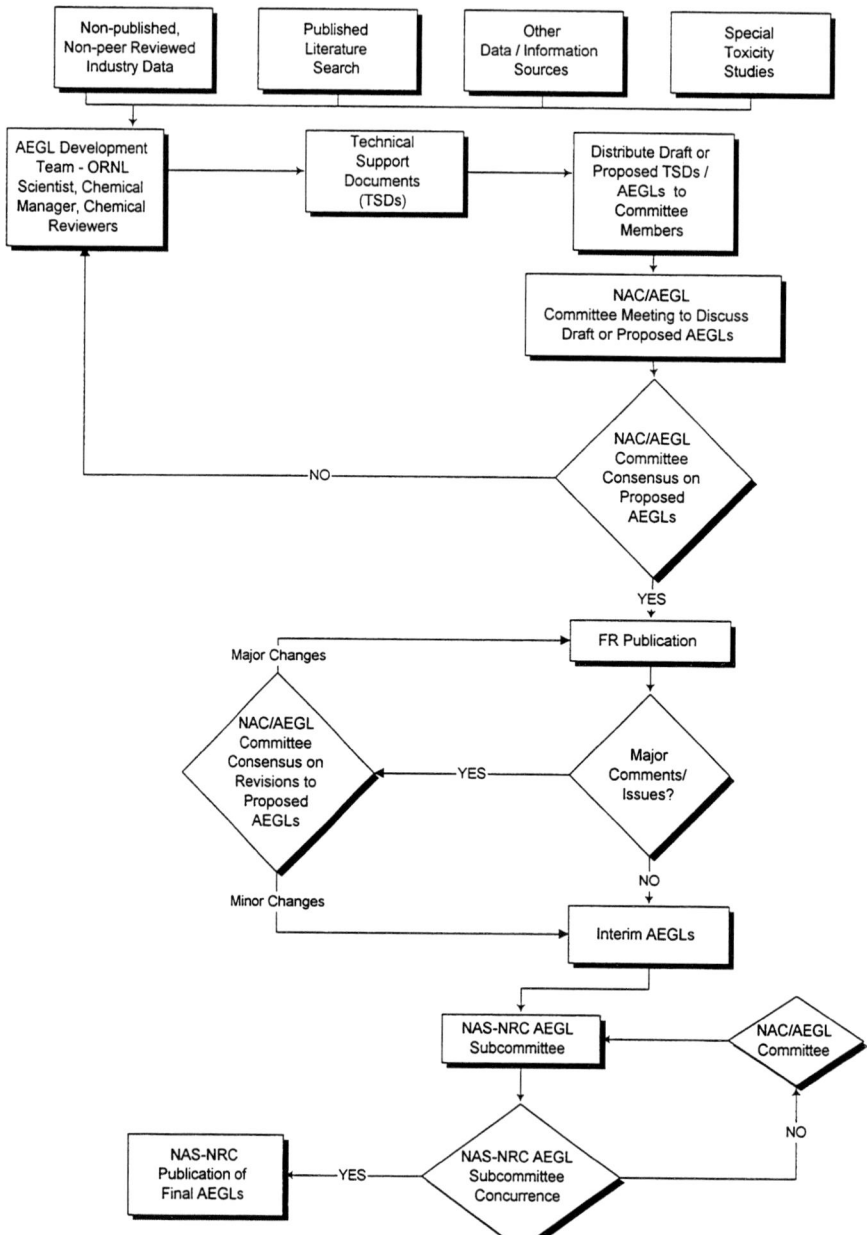

Appendix C

Glossary of Acronyms, Abbreviations, and Symbols

AAPCC	American Association of Poison Control Centers
ACGIH	American Conference of Governmental Industrial Hygienist
ACOEM	American College of Occupational and Environmental Medicine
ADI	acceptable daily intake.
AEGL	acute exposure guidelines levels
AFL-CIO	American Federation of Labor–Congress of Industrial Organizations
AIHA	American Industrial Hygienist Association
ATSDR	Agency for Toxic Substances and Disease Registry (U.S.)
BMC	benchmark concentration
BMC_{05}	benchmark concentration, 5% response
BMC_{10}	benchmark concentration, 10% response
CAAA	Clean Air Act Amendments (U.S. EPA)
CAER	Community Awareness and Emergency Response
CMA	Chemical Manufacturers Association, now known as the American Chemistry Council
CAS	Chemical Abstract Service (U.S.)
CDC	Centers for Disease Control and Prevention (U.S. DHHS)
CEEL	community emergency exposure levels (U.S. NRC)

175

CEL	continuous exposure limits (U.S. NRC)
COT	Committee on Toxicology (U.S. NRC)
C × t	concentration × time
CURE	Chemical Unit Record Estimate database
DFO	designated federal officer
DOD	Department of Defense (U.S.)
DOE	Department of Energy (U.S.)
DOT	Department of Transportation (U.S.)
DTIC	Defense Technical Information Center (U.S.)
ECETOC	European Chemical Industry Ecology and Toxicology Centre
EEGL	emergency exposure guidance levels (U.S. NRC)
EEL	emergency exposure limits (U.S. NRC)
einsatz-toleranzwert	[action tolerance levels] (Federation for the Advancement of German Fire Prevention)
EPA	Environmental Protection Agency (U.S.)
ERP	emergency response planning (AIHA)
ERPG	emergency response and planning guidelines (AIHA)
FACA	Federal Advisory Committee Act (U.S.)
FDA	Food and Drug Administration (U.S.)
FEDRIP	Federal Research in Progress (U.S.)
FEMA	Federal Emergency Management Agency (U.S.)
FEV_1	forced expiratory volume
GLP	Good Laboratory Practices
GSA	General Services Administration (U.S.)
HEAST	Health Effects Assessment Tables
HSDB	Hazardous Substances Data Base
HUD	Department of Housing and Urban Development (U.S.)
IARC	International Agency for Research on Cancer
IDLH	immediately dangerous to life and health (U.S. NIOSH)
IPCS	International Programme for Chemical Safety
IRIS	Integrated Risk Information System (U.S. EPA)
LC_{01}	lethal concentration, 1 % lethality
LC_{50}	lethal concentration, 50 % lethality
LCL	lower confidence limit
LOAEL	lowest-observed-adverse-effect level
MAC	minimum alveolar concentration
MAC	maximum acceptable concentration (The Netherlands)
MAK	maximale arbeitsplatzkonzentration [maximum workplace concentration] 8-h time-weighted average (German Research Association)

MAK. S	spitzenbegrenzung (kategorie ii, 2) [MAK peak limit (category II, 2)] 30 min × 2 per day (Germany)
MCS	multiple chemical sensitivity
MF	modifying factor
MLE	maximum likelihood estimate
MLE_{01}	maximum likelihood estimate, 1% response
MTD	maximum tolerated dose
N/A	not applicable
NAAQS	National Ambient Air Quality Standards (U.S.)
NAC	National Advisory Committee
NAC/AEGL	National Advisory Committee for Acute Exposure Guideline Levels for Hazardous Substances (NAC/AEGL Committee)
NAS	National Academy of Sciences (U.S.)
NASA	National Aeronautical and Space Administration (U.S.)
NCI	National Cancer Institute (U.S.)
NIOSH	National Institute for Occupational Safety and Health (U.S.)
NOAEL	no-observed-adverse-effect level
NRC	National Research Council (U.S.)
NRC/AEGL	National Research Council Subcommittee on Acute Exposure Guideline Levels (NRC/AEGL Subcommittee) (U.S.)
NSF	National Science Foundation (U.S.)
NTIS	National Technical Information Service (U.S.)
NTP	National Toxicology Program (U.S.)
OECD	Organization for Economic Cooperation and Development
ORNL	Oak Ridge National Laboratories (U.S.)
OSHA	Occupational Safety and Health Administration (U.S.)
OSWER	Office of Solid Waste and Emergency Response (U.S.)
PEL-TWA	permissible exposure limit–time-weighted average (U.S. OSHA)
PEL-STEL	permissible exposure limit–short-term exposure limit (U.S. OSHA)
QA	quality assurance
QC	quality control
QSARs	quantitative structure activity relationships
REL-STEL	recommended exposure limit–short-term exposure limit (U.S. NIOSH)
REL-TWA	recommended exposure limits–time-weighted average (U.S. NIOSH)
RfC	reference concentration (U.S. EPA)
RfD	reference dose (U.S. EPA)
RTECS	Registry of Toxic Effects of Chemical Substances

SARA	Superfund Amendments and Reauthorization Act (CERCLA)
SMAC	spacecraft maximum allowable concentrations
SOP	standing operating procedures
SPEGL	short-term public emergency guidance levels (U.S. NRC)
STPL	short-term public limit (U.S. NRC)
TARA	Toxicology and Risk Assessment Document List (U.S. ORNL)
TLV-STEL	Threshold Limit Value–short-term exposure limit (ACGIH)
TLV-TWA	Threshold Limit Value–time-weighted average (ACGIH)
TSD	technical support document
UF	uncertainty factor
WEEL	workplace environmental exposure level (AIHA)

>	greater than
≥	greater than or equal to
<	less than
≤	less than or equal to
%	percent
dL	deciliter
g	gram
h	hour
μm	micrometer
μg	microgram
mg	milligram
min	minute
mL	milliliter
mm	millimeter
ppb	parts per billion
ppm	parts per million
ppt	parts per trillion

Appendix D

Example of a Table of Contents of A Technical Support Document

TABLE OF CONTENTS

PREFACE 2

LIST OF TABLES 5

SUMMARY 6

1. INTRODUCTION 9
2. HUMAN TOXICITY DATA 10
   - 2.1. Acute Lethality 10
   - 2.2. Nonlethal Toxicity 10
     - 2.2.1. Acute Studies 10
     - 2.2.2. Epidemiologic Studies 11
   - 2.3. Developmental/Reproductive Toxicity 11
   - 2.4. Genotoxicity 11
   - 2.5. Carcinogenicity 11
   - 2.6. Summary 11

3. ANIMAL TOXICITY DATA 11
   3.1. Acute Lethality 12
        3.1.1. Nonhuman Primates 12
        3.1.2. Dogs 12
        3.1.3. Rats 12
        3.1.4. Mice 13
        3.1.5. Hamsters 14
   3.2. Nonlethal Toxicity 14
        3.2.1. Nonhuman Primates 14
        3.2.2. Dogs 14
        3.2.3. Rats 15
        3.2.4. Mice 15
   3.3. Developmental/Reproductive Toxicity 15
   3.4. Genotoxicity 18
   3.5. Carcinogenicity 19
   3.6. Summary 19

4. SPECIAL CONSIDERATIONS 20
   4.1. Metabolism and Disposition 20
   4.2. Mechanism of Toxicity 20
   4.3. Structure-Activity Relationships 21
   4.4. Other Relevant Information 21
        4.4.1. Species Variability 21
        4.4.2. Unique Physicochemical Properties 22
        4.4.3. Concurrent Exposure Issues 22

5. DATA ANALYSIS FOR AEGL-1 22
   5.1. Summary of Human Data Relevant to AEGL-1 22
   5.2. Summary of Animal Data Relevant to AEGL-1 22
   5.3. Derivation of AEGL-1 22

6. DATA ANALYSIS FOR AEGL-2 23
   6.1. Summary of Human Data Relevant to AEGL-2 23
   6.2. Summary of Animal Data Relevant to AEGL-2 23
   6.3. Derivation of AEGL-2 23

7. DATA ANALYSIS FOR AEGL-3 24
   7.1. Summary of Human Data Relevant to AEGL-3 24
   7.2. Summary of Animal Data Relevant to AEGL-3 24
   7.3. Derivation of AEGL-3 24

8. SUMMARY OF AEGLS 25
   8.1. AEGL Values and Toxicity Endpoints 25
   8.2. Comparison with Other Standards and Criteria 26
   8.3. Data Adequacy and Research Needs 27

9. REFERENCES CITED 29

Appendix A (Derivation of AEGL Values) 32
Appendix B (Time Scaling Calculations for Dimethylhydrazine AEGLs) 36
Appendix C (Carcinogenicity Assessment for Dimethylhydrazine) 39
Appendix D (Derivation Summary for Dimethylhydrazine AEGLs) ... 41

Appendix E

Example of a Summary of A Technical Support Document

SUMMARY

Dimethylhydrazine occurs as symmetrical (1,2-dimethylhydrazine) and unsymmetrical (1,1-dimethylhydrazine) isomers. Unless otherwise specified, dimethylhydrazine refers to unsymmetrical dimethylhydrazine in this document. Both compounds are clear, colorless liquids. 1,1-Dimethylhydrazine is a component of rocket fuels and is also used as an adsorbent for acid gas, as a plant-growth control agent, and in chemical synthesis. Although it has been evaluated as a high-energy rocket fuel, commercial use of 1,2-dimethylhydrazine is limited to small quantities, and it is usually considered to be a research chemical. Because data are limited for 1,2-dimethylhydrazine, the acute exposure guideline level (AEGL) values for both isomers are based upon 1,1-dimethylhydrazine. Limited data suggest that 1,1-dimethylhydrazine may be somewhat more toxic than 1,2-dimethylhydrazine.

Data on acute exposures of humans to both isomers of dimethylhydrazine are limited to case reports of accidental exposures. Signs and symptoms of exposure include respiratory irritation, pulmonary edema, nausea, vomiting, and neurologic effects. However, definitive exposure data (concentration and duration) were unavailable for these accidents. The limited data in humans suggest that the nonlethal toxic response to acute inhalation of dimethylhydrazine is qualitatively similar to that observed in animals. No information was available regarding lethal responses in humans. In the absence of quantitative data in humans, the use of animal data is considered a credible approach for developing AEGL values.

Toxicity data of varying degrees of completeness are available for several laboratory species, including, rhesus monkeys, dogs, rats, mice, and hamsters (Weeks et al. 1963). Most of the animal studies were conducted using 1,1-dimethylhydrazine, although limited data suggest that 1,2-dimethylhydrazine exerts similar toxic effects. Minor nonlethal effects such as respiratory tract irritation appear to occur at cumulative exposures of less than 100 parts per million multiplied by hours (ppm·h). At cumulative exposures of 100 ppm·h or slightly greater than this level, more notable effects have been reported, including, muscle fasciculation, behavioral changes, tremors, and convulsions. Lethality has been demonstrated when cumulative exposures exceed these levels only slightly. The available data suggest that there is a very narrow margin between exposures resulting in no significant toxicity and those causing substantial lethality (the lethal concentration for 50% of the animals (LC_{50}) ≈ 900-2,000 ppm·h).

Developmental toxicity of dimethylhydrazines has been demonstrated in rats following parenteral administration of maternally toxic doses.

Both isomers of dimethylhydrazine have been shown to be carcinogenic in rodents following chronic oral exposure and 6-mon inhalation exposure to 1,1-dimethylhydrazine. Increased tumor incidence was observed in mice, although these findings are compromised by the contaminant exposure to dimethylnitrosamine. An increased incidence of lung tumors and hepatocellular carcinomas was also seen in rats but not in similarly exposed hamsters. The U.S. Environmental Protection Agency (U.S. EPA) inhalation slope factors are currently unavailable for dimethylhydrazine.

AEGL-1 values for dimethylhydrazine are not recommended because of inadequate data to develop health-based criteria and because the concentration-response relationship for dimethylhydrazine indicated that a very narrow margin exists between exposures producing no toxic response and those resulting in significant toxicity.

Behavioral changes and muscle fasciculations in dogs exposed for 15 min to 1,1-dimethylhydrazine at 360 ppm (Weeks et al. 1963) served as the basis for deriving AEGL-2 values. Available lethality data in dogs and rats indicated a near linear temporal relationship (n = 0.84 and 0.80 for dogs and rats, respectively). For temporal scaling ($C^1 \times t = k$) to derive values for AEGL-specific exposure durations, a linear concentration-response relationship, n = 1, was used. (C = exposure concentration, t = exposure duration, and k = a constant.) This value was adjusted by an uncertainty factor of 30. An uncertainty factor of 3 for interspecies variability was applied, because the toxic response to dimethylhydrazine was similar across the species tested. This was especially true for lethality among rats, mice, dogs, and hamsters with LC_{50} values for time periods ranging from 5 min to 4 h. A comparison of LC_{50} values for the same exposure durations in these species did not vary more than 3-fold. An uncertainty factor of 10 was used for intraspecies variability. This factor was based primarily on the variability observed in dogs in which responses varied from one of extreme severity (vomiting, tremors, convulsions, and death) to no observable effects. Additionally, Weeks et al. (1963) indicated that dogs previously stressed by auditory stimuli may have potentiated their response to dimethylhydrazine. Based on these data, it was assumed that humans may be equally variable in their response to dimethylhydrazine as a result of similar stresses.

The AEGL-3 values were derived from the 1-h LC_{50} (981 ppm) for 1,1-dimethylhydrazine in dogs (Weeks et al. 1963). Because of the steep slope of the dose-response curve of 1,1-dimethylhydrazine, the 1-h LC_{50} of 981 ppm was adjusted to estimate the lethality threshold of 327 ppm. An uncertainty factor of 3 for interspecies variability was applied for several reasons. The 4-h LC_{50} values for mouse, rat, and hamster differ by a factor of approximately 2 and were consistent with the dog data when extrapolated from 1 h using n = 1. The more susceptible species, the dog, was used to derive the AEGL-3 values. An uncertainty factor of 10 for intraspecies variability was used because a broad spectrum of effects were seen including behavioral effects, hyperactivity, fasciculations, tremors, convulsions, and vomiting. The mechanism of toxicity is uncertain, and susceptibility among individuals may vary. Following identical exposures, the responses of the dogs varied from one of extreme severity (vomiting, tremors, convulsions, and death) to no observable effects. Temporal scaling as previously described was applied to obtain exposure values for AEGL-specific exposure periods.

Verified inhalation and oral slope factors were unavailable from U.S. EPA for dimethylhydrazine. A cancer assessment based upon the carcinogenic potential (withdrawn cancer slope factors) of dimethylhydrazine revealed that AEGL values for a theoretical excess lifetime 10^{-4} carcinogenic risk exceeded the AEGL-2 values that were based on noncancer endpoints. Because the risk for dimethylhydrazine exposure was estimated from nonverified sources and because AEGLs are applicable to rare events or single once-in-a-lifetime exposures to a limited geographic area and small population, the AEGL values based on noncarcinogenic endpoints were considered to be more appropriate.

Summary of AEGL Values for 1,1- and 1,2-Dimethylhydrazines

Classification	30 min	1 h	4 h	8 h	Endpoint (Reference)
AEGL-1 (Nondisabling)	NR	NR	NR	NR	Not recommended due to insufficient data; concentration-response relationships suggest little margin between exposures causing minor effects and those resulting in serious toxicity[a]
AEGL-2 (Disabling)	6 ppm (14.7 mg/m^3)	3 ppm (7.4 mg/m^3)	0.75 ppm (2 mg/m^3)	0.38 ppm (1 mg/m^3)	Behavioral changes and muscle fasciculations in dogs exposed at 360 ppm for 15 min (Weeks et al. 1963)
AEGL-3 (Lethal)	22 ppm (54 mg/m^3)	11 ppm (27 mg/m^3)	2.7 ppm (6.6 mg/m^3)	1.4 ppm (3.4 mg/m^3)	Lethality threshold of 327 ppm for 1 h estimated from 1-h LC$_{50}$ in dogs (Weeks et al. 1963)

Numeric values for AEGL-1 are not recommended because (1) available data are lacking, (2) data indicate that toxic effects may occur at or below the odor threshold, (3) the margin of safety that exists between the derived AEGL-1 and the AEGL-2 is inadequate, or (4) the derived AEGL-1 is greater than the AEGL-2. Absence of an AEGL-1 does not imply that exposure below the AEGL-2 is without adverse effects.

Abbreviations: NR, not recommended; ppm, parts per million; mg/m^3, milligrams per cubic meter.

Reference: Weeks, M.H., Maxey, G.C., Sicks, Greene, E.A. 1963. Vapor toxicity of UDMH in rats and dogs from short exposures. *American Industrial Hygiene Association Journal* 24:137-143.

Appendix F

Example of the Derivation of AEGL Values Appendix in A Technical Support Document

DERIVATION OF AEGL-1 VALUES

Key study: None. An AEGL-1 was not recommended because of inadequate data for developing health-based criteria and because exposure-response relationships suggest little margin between exposures resulting in no observable adverse effects and those producing significant toxicity. The absence of an AEGL-1 does not imply that exposure below the AEGL-2 is without adverse effects.

DERIVATION OF AEGL-2 VALUES

Key study: Weeks et al. 1963

Toxicity endpoint: Dogs exposed to 1,1-dimethylhydrazine at 360 ppm for 15 min exhibited behavioral changes and muscle fasciculations

Uncertainty factors:	An uncertainty factor of 3 for interspecies variability was applied because the toxic response to dimethylhydrazine was similar across the species tested. This was especially true for lethality responses (LC_{50} values for varying time periods ranging from 5 min to 4 h) among rats, mice, dogs, and hamsters. A comparison of LC_{50} values for the same exposure durations in these species did not vary more than 3-fold. An uncertainty factor of 10 was retained for intraspecies variability (protection of sensitive populations). A broad spectrum of effects were seen that included behavioral effects, hyperactivity, fasciculations, tremors, convulsions, and vomiting. The mechanism of toxicity is uncertain and susceptibility among individuals regarding these effects may vary. Following identical exposures, the responses of the dogs varied from extreme severity (vomiting, tremors, convulsions, and death) to no observable effects. A factor of 10 was also applied because experiments by Weeks et al. (1963) indicated that dogs that had been previously stressed (auditory stimuli) were more susceptible to the adverse effects of dimethylhydrazine.
Calculations:	360 ppm/30 = 12 ppm $C^1 \times t = k$ 12 ppm × 15 min = 180 ppm·min
Time scaling:	$C^1 \times t = k$ (ten Berge et al. 1986) $(12 \text{ ppm})^1 \times 15 \text{ min} = 180 \text{ ppm·min}$ LC_{50} data were available for 5-, 15-, 30-, 60-, and 240-min exposures in rats and 5, 15, and 60 min in dogs. Exposure-response data indicated a near linear concentration-response relationship (n = 0.84 for rats; n = 0.80 for dogs). For time-scaling, a linear relationship was assumed and a value of n = 1 was selected.

30-min AEGL-2: $C^1 \times 30 \text{ min} = 180 \text{ ppm·min}$
$C = 6$ ppm
1-h AEGL-2: $C^1 \times 60 \text{ min} = 180 \text{ ppm·min}$
$C = 3$ ppm
4-h AEGL-2: $C^1 \times 240 \text{ min} = 180 \text{ ppm·min}$

8-h AEGL-2:
C = 0.75 ppm
$C^1 \times 480$ min = 180 ppm·min
C = 0.38 ppm

DERIVATION OF AEGL-3

Key study: Weeks et al. 1963

Toxicity endpoint: 1-h LC_{50} of 981 ppm in dogs reduced by a factor of three to 327 ppm as an estimate of a lethality threshold. Weeks et al. (1963) provided data showing that 15-min exposure of dogs at 36-400 ppm produced only minor, reversible effects (behavioral changes and mild muscle fasciculations)

Uncertainty factors: An uncertainty factor of 3 for interspecies variability was applied because the toxic response to dimethylhydrazine was similar across the species tested. This was especially true for lethality responses (LC_{50} values for varying time periods ranging from 5 min to 4 h) among rats, mice, dogs, and hamsters. A comparison of LC_{50} values for the same exposure durations in these species did not vary more than 3-fold.
An uncertainty factor of 10 was applied for intraspecies variability (protection of sensitive populations). A broad spectrum of effects were seen that included behavioral effects, hyperactivity, fasciculations, tremors, convulsions, and vomiting. The mechanism of toxicity is uncertain and susceptibility among individuals regarding these effects may vary. Following identical exposures, the responses of the dogs varied from extreme severity (vomiting, tremors, convulsions, and death) to no observable effects. A factor of 10 was also applied because experiments by Weeks et al. (1963) indicated that dogs that had been previously stressed (auditory stimuli) were more susceptible to the adverse effects of dimethylhydrazine.

Calculations: 327 ppm/30 = 10.9 ppm
$C^1 \times t = k$
11.9 ppm × 60 min = 654 ppm·min

Time scaling: $C^1 \times t = k$ (ten Berge et al.1986)
11.9 ppm^1 × 60 min = 654 ppm·min

LC$_{50}$ data were available for 5, 15, 30, 60, and 240-min exposures in rats and 5, 15, and 60 min in dogs. Exposure-response data indicated a near linear concentration-response relationship (n = 0.84 for rats, n = 0.80 for dogs). For time-scaling, a linear relationship was assumed and a value of n = 1 was selected.

30-min AEGL-2: C^1 × 30 min = 654 ppm·min
C = 22 ppm
1-h AEGL-2: C^1 × 60 min = 654 ppm·min
C = 11 ppm
4-h AEGL-2: C^1 × 240 min = 654 ppm·min
C = 2.7 ppm
8-h AEGL-2: C^1 × 480 min = 654 ppm·min
C = 1.4 ppm

Appendix G

Example of Time-Scaling Calculations Appendix in A Technical Support Document

TIME-SCALING CALCULATIONS FOR DIMETHYLHYDRAZINE AEGLs

The relationship between dose and exposure time to produce a toxic effect for any given chemical is a function of the physical and chemical properties of the substance and the unique toxicologic and pharmacologic properties of the individual substance. Historically, the relationship according to Haber (1924), commonly called Haber's law (NRC 1993) or Haber's rule (i.e., $C \times t = k$, where C = exposure concentration, t = exposure duration, and k = a constant) has been used to relate exposure concentration and duration to a toxic effect (Rinehart and Hatch 1964). This concept states that exposure concentration and exposure duration may be reciprocally adjusted to maintain a cumulative exposure constant (k) and that this cumulative exposure constant will always reflect a specific quantitative and qualitative response. This inverse relationship of concentration and time may be valid when the toxic response to a chemical is equally dependent upon the concentration and the exposure duration. However, an assessment by ten Berge et al. (1986) of LC_{50} data for certain chemicals revealed chemical-specific relationships between exposure concentration and exposure duration that were often exponential. This relationship can be expressed by the equation $C^n \times t = k$, where n represents a chemical-specific and even a toxic endpoint-specific exponent. The relationship described by this equation is basically the form of a linear regression analysis of the log-log transformation of a plot of C vs t. ten Berge et al. (1986) examined the airborne concentration (C) and short-term exposure duration (t) relationship relative to death for approximately 20 chemicals and found that the empirically derived value of n ranged from 0.8 to 3.5 among this group of chemicals. Hence, these workers showed that the value of the exponent (n) in the equation $C^n \times t = k$ quantitatively defines the relationship between exposure concentration and exposure duration for a given chemical and for a specific health effect endpoint. Haber's rule is the special case where n = 1. As the value of n increases, the plot of concentration vs time yields a progressive decrease in the slope of the curve.

Two data sets of LC_{50} values for different time periods of exposure were analyzed using a linear regression analysis of the log-log transformation of a plot of C vs t to derive values of n for dimethylhydrazine.

Dimethylhydrazine Dog Data from Weeks et al. 1963

The LC_{50} values for 5-, 15-, and 60-min exposures were 22,300, 3,580, and 981 ppm, respectively.

Time	Concentration	Log Time	Log Concentration
5	22,300	0.6990	4.3483
15	3,580	1.1761	3.5539
60	981	1.7782	2.9917

n = 0.8

Calculated LC_{50} values:

Min	Concentration
30	2036.15
60	860.12
240	153.48
480	64.83

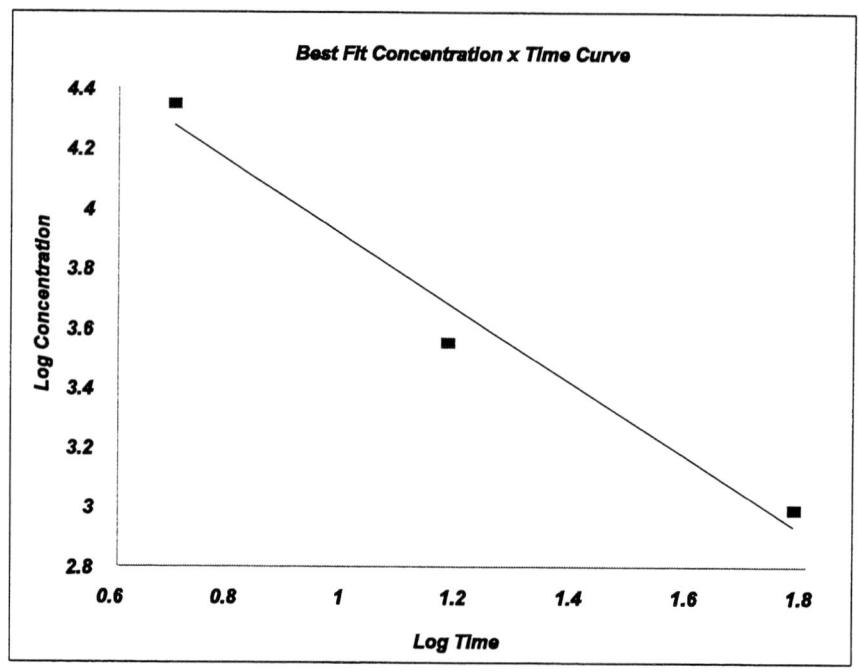

Dimethylhydrazine Rat Data from Weeks et al. 1963

The LC_{50} values for 5-, 15-, 30-, 60-, and 240-min exposures were 24,500, 8,230, 4,010, 1,410, and 252 ppm, respectively.

Time	Concentration	Log Time	Log Concentration
5	24,500	0.6990	4.3892
15	8,230	1.1761	3.9154
60	4,010	1.4771	3.6031
240	252	2.3802	2.4014

n = 0.84

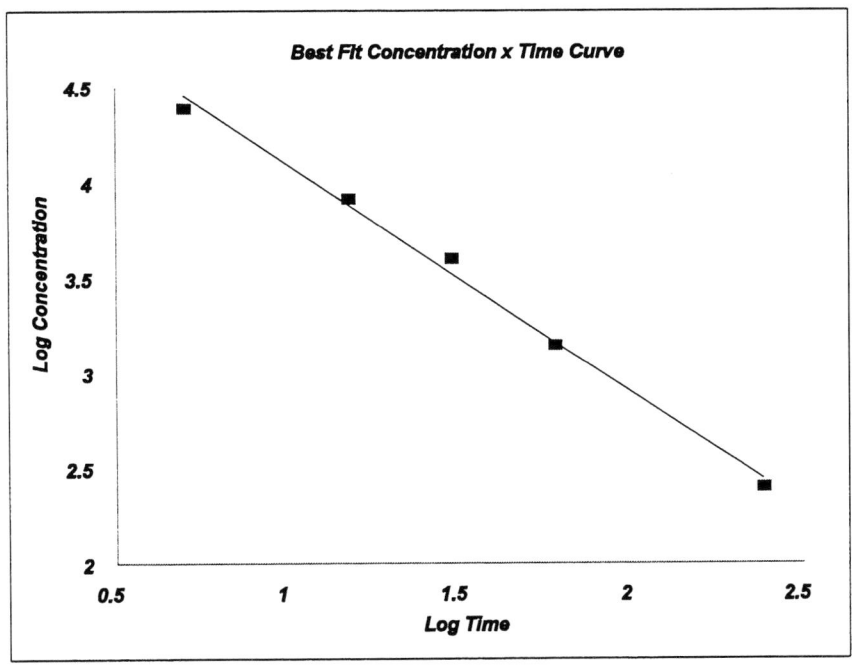

Calculated LC_{50} values:

Min	Concentration
30	3,323.28
60	1,449.93
240	276.00
480	120.42

Appendix H

Example of a Carcinogenicity Assessment Appendix in A Technical Support Document

CARCINOGENICITY ASSESSMENT OF DIMETHYLHYDRAZINE

Slope factors for 1,1-dimethylhydrazine and 1,2-dimethylhydrazine were available but have been withdrawn from the U.S. EPA Integrated Risk Information System (IRIS) (U.S. EPA 1986). For a preliminary carcinogenicity assessment, the withdrawn inhalation slope factor for 1,1-dimethylhydrazine (cited in ATSDR 1994) will be used. The assessment follows previously described methodologies (NRC 1985; Henderson 1992).

The withdrawn slope factor for 1,1-dimethylhydrazine was 3.5 $(mg/kg \cdot d)^{-1}$, which, based upon a human inhalation rate of 20 m^3/d and a body weight of 70 kg, is equivalent to 1 $(mg/m^3)^{-1}$.

To convert to a level of monomethylhydrazine that would cause a theoretical excess cancer risk of 10^{-4}:

Risk of $1 \times 10^{-4} = (1 \times 10^{-4}/1) \times 1 \text{ mg/m}^3 \quad = \quad 1 \times 10^{-4} \text{ mg/m}^3$
(virtually safe dose)

To convert a 70-y exposure to a 24-h exposure:

24-h exposure = d × 25,600
            = $(1 \times 10^{-4} \text{ mg/m}^3) \times 25,600$ d
            = 2.56 mg/m^3

To account for uncertainty regarding the variability in the stage of the cancer process at which monomethylhydrazine or its metabolites may act, a multistage factor of 6 is applied (Crump and Howe 1984):

(2.56 mg/m^3)/6 = 0.43 mg/m^3 (0.18 ppm)

Therefore, based upon the potential carcinogenicity of monomethylhydrazine, an acceptable 24-h exposure would be 0.9 mg/m^3 (0.5 ppm).

If the exposure is limited to a fraction (f) of a 24-h period, the fractional exposure becomes 1/f × 24 h (NRC 1985).

24-h exposure	=	0.43 mg/m^3 (0.18 ppm)
8-h	=	1.3 mg/m^3 (0.5 ppm)
4-h	=	2.6 mg/m^3 (1.1 ppm)
1-h	=	10.3 mg/m^3 (4.2 ppm)
0.5 h	=	20.6 mg/m^3 (8.5 ppm)

Because the AEGL-2 values based upon acute toxicity were equivalent to or lower than the 10^{-4} risk values derived based on potential carcinogenicity, the acute toxicity data were used for the AEGLs for dimethylhydrazine. For 10^{-5} and 10^{-6} risk levels, the 10^{-4} values are reduced by 10-fold or 100-fold, respectively.

Appendix I

Example of the AEGL Derivation Summary Appendix in A Technical Support Document

**DERIVATION SUMMARY FOR
ACUTE EXPOSURE GUIDELINE LEVELS
FOR DIMETHYLHYDRAZINE
(CAS No. 57-14-7; 1,1-Dimethylhydrazine)
(CAS No. 540-73-8; 1,2-Dimethylhydrazine)**

AEGL-1 Values			
30 min	1 h	4 h	8 h
Not recommended	Not recommended	Not recommended	Not recommended
Reference: Not applicable			
Test Species/Strain/Number: Not applicable			
Exposure Route/Concentrations/Durations: Not applicable			
Effects: Not applicable			
Endpoint/Concentration/Rationale: Not applicable			*(Continued)*

Uncertainty Factors/Rationale: Not applicable
Modifying Factor: Not applicable
Animal to Human Dosimetric Adjustment: Not applicable
Time Scaling: Not applicable
Data Adequacy: Numeric values for AEGL-1 are not recommended because (1) data are not available, (2) data indicate that toxic effects may occur at or below the odor threshold, (3) an inadequate margin of safety exists between the derived AEGL-1 and the AEGL-2, or (4) the derived AEGL-1 is greater than the AEGL-2. Absence of an AEGL-1 does not imply that exposure below the AEGL-2 is safe.

NOTE: If an AEGL-1 value is not recommended, there should be a short discussion of the rationale for that choice. The rationale should include as appropriate a discussion that numeric values for AEGL-1 are not recommended because (1) relevant data are lacking, (2) the margin of safety between the derived AEGL-1 and AEGL-2 values is inadequate, or (3) the derived AEGL-1 is greater than the AEGL-2. Absence of an AEGL-1 does not imply that exposure below the AEGL-2 is safe.

AEGL-2 Values			
30 min	1 h	4 h	8 h
6.0 ppm	3.0 ppm	0.75 ppm	0.38 ppm
Reference: Weeks, M.H., G.C. Maxey, M.E. Sicks, and E.A. Greene. 1963. Vapor toxicity on UDMH in rats and dogs from short exposures. Am. Ind. Hyg. Assoc. J. 24:137-143			
Test Species/Strain/Sex/Number: mongrel dogs, 2-4/group, sex not specified			
Exposure Route/Concentrations/Durations: Inhalation; 1,200-4,230 ppm for 5 min; 360, 400, or 1,530 ppm for 15 min; 80-250 ppm for 60 min			

Effects:	
Exposure (15 min)	Effect
360 ppm	muscle fasciculations in 1 of 4 dogs (determinant for AEGL-2)
400 ppm	behavioral changes in 2 of 4 dogs
1,530 ppm	tremors, convulsions, vomiting in 2 of 2 dogs

Endpoint/Concentration/Rationale: 15-min exposure at 360 ppm considered a threshold for potentially irreversible effects or effects that would impair escape. At this exposure, muscle fasciculations were observed in 1 of 4 exposed dogs, and at 400 ppm, behavioral changes were observed. *(Continued)*

Uncertainty Factors/Rationale: Total uncertainty factor: 30 Interspecies: 3 - The toxic response to dimethylhydrazine (LC_{50} values) was similar across species. The 4-h LC_{50} values for mouse, rat, and hamster differ by a factor of approximately 2 and were consistent with the dog data when extrapolated from 1 h using n = 1. The more susceptible species, the dog, was used to derive the AEGL-2 values. Intraspecies: 10 - A broad spectrum of effects were seen, including behavioral effects, hyperactivity, fasciculations, tremors, convulsions, and vomiting. The mechanism of toxicity is uncertain and susceptibility among individuals regarding these effects may vary. This variability was especially demonstrated in dogs wherein responses varied from one of extreme severity (vomiting, tremors, convulsions, and death) to no observable effects. Therefore, a factor of 10 was applied. A factor of 10 was also applied because experiments by Weeks et al. (1963) indicated that dogs had been previously stressed (auditory stimuli), which may have affected their response to dimethylhydrazine. Based upon these data, it was assumed that humans may be equally variable in their response to dimethylhydrazine.
Modifying Factor: None
Animal to Human Dosimetric Adjustment: None applied, insufficient data
Time Scaling: $C^n \times t = k$, where n = 1 and k = 180 ppm·min; LC_{50} data were available for 5-, 15-, 30-, 60-, and 240-min exposures in rats and 5-, 15-, and 60-min in dogs. Exposure-response data indicated a near linear concentration-response relationship (n = 0.84 for rats; n = 0.80 for dogs). For time-scaling, a linear relationship was assumed and a value where n = 1 was selected.
Data Adequacy: Information regarding the human experience for acute inhalation exposure to dimethylhydrazine are limited to qualitatively case reports indicating nasal and respiratory tract irritation, breathing difficulties, and nausea. Data in animals have shown concentration-dependent effects ranging from respiratory tract irritation, pulmonary edema and neurologic effects to lethality. Because the nonlethal effects in humans and animals are qualitatively similar, the animal data were considered relevant and appropriate for development of AEGL values. The AEGL values for dimethylhydrazine reflect the steep exposure-response relationship suggested by available data.

AEGL-3 Values			
30 min	1 h	4 h	8 h
22 ppm	11 ppm	2.7 ppm	1.4 ppm

Reference: Weeks, M.H., G.C. Maxey, M.E. Sicks, and E.A. Greene. 1963. Vapor toxicity of UDMH in rats and dogs from short exposures. Am. Ind. Hyg. Assoc. J. 24:137-143

Test Species/Strain/Sex/Number: mongrel dogs, 3-4/group; sex not specified

Exposure Route/Concentrations/Durations: Inhalation; exposure to various concentrations (80-22,300 ppm) for 5, 15, or 60 min

Effects:
  1-h LC_{50} 981 ppm (reduction by 1/3 was basis for AEGL-3 derivation)
  15-min LC_{50} 3,580 ppm
  5-min LC_{50} 22,300 ppm

Endpoint/Concentration/Rationale: 1-h LC_{50} (981 ppm) reduced by 1/3 was considered an estimate of the lethality threshold (327 ppm). Based on the available exposure-response data for this chemical (Jacobson et al. 1955), a 3-fold reduction in LC_{50} values results in exposures that would not be associated with lethality.

Uncertainty Factors/Rationale: Total uncertainty factor: 30
  Interspecies: 3 - The toxic response to dimethylhydrazine (LC_{50} values) was similar across species. The 4-h LC_{50} values for mouse, rat, and hamster differ by a factor of approximately 2 and were consistent with the dog data when extrapolated from 1 h using n = 1. The more susceptible species, the dog, was used to derive the AEGL-3 values.
  Intraspecies: 10 - A broad spectrum of effects were seen, including behavioral effects, hyperactivity, fasciculations, tremors, convulsions, and vomiting. The mechanism of toxicity is uncertain, and susceptibility among individuals regarding these effects may vary. This variability was especially demonstrated in dogs wherein responses varied from one of extreme severity (vomiting, tremors, convulsions, and death) to no observable effects. Therefore, a factor of 10 was used. A factor of 10-fold was also used because experiments by Weeks et al. (1963) indicated that dogs previously stressed by auditory stimuli may have a potentiated response to dimethylhydrazine. Based upon these data, it was assumed that humans may be equally variable in their response to dimethylhydrazine subsequent to similar stresses.

Modifying Factor: None

Animal to Human Dosimetric Adjustment: None applied, insufficient data

Time Scaling: $C^n \times t = k$, where $n = 1$ and $k = 654$ ppm·min; LC_{50} data were available for 5-, 15-, 30-, 60-, and 240-min exposures in rats and 5-, 15-, and 60-min in dogs. Exposure-response data indicated a near linear concentration-response relationship ($n = 0.84$ for rats; $n = 0.80$ for dogs). For time-scaling, a linear relationship was assumed and a value where $n = 1$ was selected by the National Advisory Committee.

Data Adequacy: Information regarding the lethality of dimethylhydrazine in humans were not available. Lethality data for several animal species allowed for a defensible development of the AEGL-3 values but uncertainties remain regarding individual variability in the toxic response to dimethylhydrazines.

Appendix J

List of Extant Standards and Guidelines in A Technical Support Document

Section 8.2 of the technical support document (TSD) compares the AEGL values for a chemical with other standards and guidelines previously published for exposure durations ranging from 10 min to 8 h. A summary discussion of important comparisons should be presented in the text and the values for recognized standards and guidelines, if available, should be presented in the table. The statement, "All currently available standards and guidelines are shown in Table . . ." should be included in the text to affirm completeness of the table. Only those standards or guidelines with published values for a given chemical should be included in the table. In cases where the exposure duration of a published standard or guideline differs from those designated for AEGLs (e.g., 15-min PEL-STEL), the value should be placed in parentheses in the column of the closest AEGL exposure duration category and footnoted to indicate its true exposure duration. A list of recognized standards and guidelines and the order in which they should appear in the following table.

LIST AND ORDER OF PRESENTATION OF EXTANT STANDARDS AND GUIDELINES IN THE TSD TABLE

AEGL-1	acute exposure guideline level-1
AEGL-2	acute exposure guideline level-2
AEGL-3	acute exposure guideline level-3
ERPG-1 (AIHA)	emergency response planning guideline-level 1
ERPG-2 (AIHA)	emergency response planning guideline-level 2
ERPG-3 (AIHA)	emergency response planning guideline-level 3
SPEGL (NRC)	short-term emergency guidance level
EEL (NRC)	emergency exposure limit
STPL (NRC)	short-term public limit
CEL (NRC)	continuous exposure limit
EEGL (NRC)	emergency exposure guidance level
SMAC (NRC)	spacecraft maximum allowable concentration for space-station contaminants
PEL-STEL (OSHA)	permissible exposure limit–short-term exposure limit
PEL-TWA (OSHA)	permissible exposure limit–time weighted average
IDLH (NIOSH)	immediately dangerous to life and health
REL-STEL (NIOSH)	recommended exposure limit
TLV-STEL (ACGIH)	Threshold Limit Value–short-term exposure limit
TLV-TWA (ACGIH)	Threshold Limit Value–time-weighted average
MAC (Netherlands)	maximum acceptable concentration
MAK (Germany)	[maximale arbeitsplatzkonzentration] maximum workplace concentration, 8-h time weighted average, German Research Association
MAK S. (Germany)	spitzenbegrenzung (kategorie ii, 2) [peak limit II, 2] 30 min × 2 per day
einsatztoleranzwert (Germany)	action tolerance levels